DIMENSIONS OF HEALTH AND HEALTH PROMOTION

AF840999

VIBS

Volume 138

Robert Ginsberg
Founding Editor

Peter A. Redpath
Executive Editor

Associate Editors

G. John M. Abbarno
Mary-Rose Barral
Gerhold K. Becker
Raymond Angelo Belliotti
Kenneth A. Bryson
C. Stephen Byrum
H. G. Callaway
Robert A. Delfino
Rem B. Edwards
William Gay
Dane R. Gordon
J. Everet Green
Heta Aleksandra Gylling
Matti Häyry
Steven V. Hicks
Richard T. Hull
Laura Duhan Kaplan
Joseph C. Kunkel
Vincent L. Luizzi
Alan Milchman
George David Miller
Jon Mills
Alan Rosenberg
Arleen L. F. Salles
Steven Schroeder
John R. Shook
Alan Soble
Eddy Souffrant
Tuija Takala
Oscar Vilarroya
Anne Waters
John R. Welch

DIMENSIONS OF HEALTH AND HEALTH PROMOTION

Edited by

Lennart Nordenfelt
Per-Erik Liss

Rodopi

Amsterdam - New York, NY 2003

The paper on which this book is printed meets the requirements of "ISO 9706:1994, Information and documentation - Paper for documents - Requirements for permanence".

ISBN: 90-420-0924-1
©Editions Rodopi B.V., Amsterdam - New York, NY 2003
Printed in the Netherlands

CONTENTS

FOREWORD

For several years I had the pleasure and privilege of following at close range the work of the Department of Health and Society at Linköping University. I participated as a partner in many dialogues with research students and as a participant in advanced seminars.

My engagement was as a philosopher. Therefore, my main concern was with the philosophical section of the Department. Its genuine interdisciplinary approach impressed me in several ways: Old and new problems appeared together; in this process the need to promote variety in choice of subject, style, and research tradition became apparent; and, undoubtedly, the researchers worked hard to respond to that need. They have now made some of this work available in the form of a selection of essays.

I hope that this book will give readers reasons to reflect on conceptual matters in the field of health conceptions and health-promotion. One of my reflections throughout has been that doing what is sometimes misleadingly called applied philosophy is a challenging task that can yield interesting results. I also hope that the book will give practitioners in the field reasons to reflect on the advantages of exploring different approaches to topics of abiding concern to them.

Ingmar Pörn
Emeritus Professor of Philosophy
University of Helsinki

EDITORS' PREFACE

The Department of Health and Society (Tema Hälsa och samhälle, Tema H) at Linköping University, Sweden, started in 1981. The idea behind the Tema-Institute, of which Tema H formed a part, was to conjoin representatives from different disciplines to focus their research and teachings on specific areas ("temas"). Health and health care were the subjects the Institute chose for Tema H. This became the research platform for a small faculty of economists, sociologists, philosophers, historians, and an epidemiologist interested in health matters.

This book reflects some concerns and research results of Tema H's philosophical section. With one exception, the authors in this book are or have been teachers or graduate students at Tema H. The exception, Dr. Olle Hellström, is a primary care doctor who has taken part in many seminars and other academic activities in the department.

Our references to the philosophical section of Tema H are a simplification. Since the Department is interdisciplinary, no sharp borders exist between the subjects. Some authors in this volume have no academic philosophical training. Our university backgrounds range mostly from medicine, psychology, sociology, and religion to philosophy. Binding us together is our deep interest in the theory of medicine and the pursuit of a philosophy of humanistic medicine and health care.

A quick glance at the contents of this book shows that our main concern is the clarification of central concepts related to health care, especially the concept of health. This concern quite well reflects the general focus of the Tema H research. Since the publication of Lennart Nordenfelt's *On the Nature of Health* (1987, 2nd edit., 1995), several dissertations in the department have dealt with the notion of health, be it somatic or mental: Tommy Svensson, *On the Notion of Mental Illness* (1990); Per-Erik Liss, *Health Care Need* (1990); Einar Jakobsson, *Psykoterapins uppgift* [*The Aim of Psychotherapy*] (1994); Per-Anders Tengland, *Mental Health: A Philosophical Analysis* (1998, 2002); Fredrik Svenaeus, *The Hermeneutics of Medicine and the Phenomenology of Health* (1999, 2001); and Jennifer Bullington, *A New Look at Psychosomatics* (1999). Differences exist between our detailed analyses of health. We come together, however, in a common holistic outlook on the concept and in setting the goal of medicine on the level of the capability and well-being of the whole human being.

Some of the treatises published at Tema H emphasize aspects of the theory of health outlined by Nordenfelt in 1987. This theory has a place in the tradition of the American Caroline Whitbeck, the Finn Ingmar Pörn, and the Englishman K. W. M. Fulford. A central idea in this tradition is that health consists of a special kind of ability; in Nordenfelt's terms, an ability to realize the subject's vital goals given standard or accepted circumstances. Diseases, injuries, and defects are viewed as the typical causes of a reduction of the ability that constitutes health. Liss explicitly bases his analysis of health care need on this characterization of health, and Tengland develops a theory of mental health on the same ground. Other authors in the department have discussed this notion in their treatises, but have had other foci (Svensson) or chosen to develop ideas in a slightly different direction (Svenaeus). In this volume, the concept of health is the primary concern for Nordenfelt, Sundström, Bullington, and, from a historical point of view, Petersson.

Also, some negative medical concepts, such as illness, disease, and handicap, have attracted the academic interest of many of us. Sundström's *Icons of Disease* (1987) is a thorough analysis of central concepts of disease in the spirit of Paul Ricoeur. Mental illness comes to the fore in Svensson's treatment. Psychosomatic illness is the focus in Bullington's dissertation. Disease and illness and the relation between the two, from the point of view of the clinician, constitute the main topic in Hellström's writings on the philosophy of medicine. Svenaeus, in this volume, pursues an analysis of the relation between the notions of handicap and illness.

We might note that some of the sociologists of Tema H—in particular Bengt Richt and Gunilla Tegern—are pursuing research on lay conceptions of health and illness partly using the philosophical models as tools for analysis and paradigms for comparison. Richt's contribution to this volume is an example of the sociologist's concern.

The philosophical discussion of health promotion, quality of life, priority-setting, and the setting of goals for medicine and health care are other central topics at Tema H. They came to the fore when Tema H, in 1993, became part of the broad international research program called The Goals of Medicine. The director of the Hastings Center for Ethics in New York, Daniel Callahan, gathered representatives from 15 countries to consider issues concerning the goals and priorities in medicine. One central issue in these considerations was how to characterize the nature of health and quality of life—and their place as a goal in medicine. The main result of this program was the publication of the book, *The Goals of Medicine: The Forgotten Issues in Health Care Reform* (1999) to which two of the present authors contributed. A further event that triggered research and philosophical considerations in this area was the Swedish governmental report—and later a law—on prioritization in health care. Members of Tema H have been active in responding to these documents. The notion of quality of life and its relation to health and health care is an issue treated in Nordenfelt's *Concepts and Measurement of Quality of Life in Health Care* (1994) and in Ragnar Furenhed's dissertation, *En Gåtfull Verklighet* [*Puzzling Reality*], (1997). In the present volume, our interest in health promotion, quality of life, and goal-setting is most clearly visible in the contributions by Liss, Jakobsson, Zeiler, and Tengland.

Some of the texts produced by authors in the Tema H group explicitly or implicitly deal with issues in the philosophy of medical science and health care. Many of our conceptual analyses have implications for the science and the practice of medicine. We have highlighted questions of scientific methodology and requirements to make with regard to the scientific support of medical treatments. Reflections on these issues come to the fore in quite different ways in Henrik Belfrage's *Psykiskt störda brottslingar* [*Mentally Disordered Offenders*] (1989) and Elisabeth Wärnberg Gerdin's *Nordisk samhällsodontologi* [*Nordic Community Dentistry*] (1994). In this volume, Ingemar Nordin's article on quackery reflects this concern.

Questions of ethics, in particular the issue of what characterizes the good practice of medicine, have always been present, although not always the primary focus, in our writings. Svenaeus analyzes the concept of the good clinical meeting in his dissertation and in later works. All of Hellström's writings and Svenaeus's *Sjukdomens mening* [*The Meaning of Disease*] discuss the ethical requirements on the clinician, not least in view of a holistic concept of health. Sundström is the author of a textbook in medical ethics, *Sjukvårdens etiska grunder* [*The Ethical Foundations of Health Care*]. The ethics of health promotion is the topic of Nina Nikku's *Informative Paternalism* (1997); ethical and legal aspects of abortion are treated in Jan Perselli's *Fostersyn i svensk rätt* [*The Perception of the Fetus in Swedish Law*]; and the ethics of the new technologies of reproduction is the issue in Kristin Zeiler's forthcoming dissertation. In this volume, the ethical concern is salient in the contributions by Hellström, Nikku, and Zeiler.

The department harbors a variety of methods and styles, a feature that is quite apparent in the present volume. Traditions from theoretical psychology, sociology, linguistics, and religious studies mix with different philosophical traditions. Although most contributions are highly influenced by Anglo-American analytical philosophy, some texts rest on a salient phenomenological platform (Svenaeus and Bullington). To this comes the rich source of case studies from family medicine presented by Hellström.

We have published this book for two reasons. First, we wish to celebrate that Tema H has existed for 20 years and that it has been a fruitful setting for a broad, cooperative scholarly discussion on the philosophy of medicine and health care. Second, we are now at a stage when several of us will leave (or have left) the department for academic tasks in other institutions in Scandinavia. It is now timely to write a legacy of our communal efforts.

Lennart Nordenfelt and Per-Erik Liss
Linköping
July, 2002

Part One

CHARACTERIZATIONS OF HEALTH AND THE GOOD LIFE

One

HEALTH, DOCTORS, AND THE GOOD LIFE: A FOOTNOTE TO PLATO

Bo Petersson

1. Introduction

In most of his dialogues, Plato refers to the special knowledge and skills of different experts. He often refers to artisans like house-builders, carpenters, and shoemakers, but he appears to take a special interest in the work of medical doctors. The reason must be that he considers the doctor's aims and methods analogous to those of philosophers, judges, and true political leaders. In a way, they all engage in curing and caring, even if they work in different fields. Philosophers, judges, and true political leaders are a special kind of doctors because they bring about virtue in individuals and society. And, as Plato says, "Virtue seems, then, to be a kind of health, fine condition, and well-being of the soul, while vice is disease, shameful condition, and weakness" (*Republic*, 444d).

In this article, I will introduce Plato's opinions of the doctor's competence, methods, and tasks. Roughly, the aim of the doctor's work is to bring about health. Therefore, I will begin with a characterization of Plato's concept of health. As Plato often takes pleasure and desire to be threats against health, I will in the final part of my article comment on his views of pleasure and of the connection between pleasure and a good human life. My purpose, thus, is mainly descriptive.

Plato wrote his dialogues within a span of about half a century. Naturally, his opinions on a given subject can differ from text to text and from one time to another. As his texts are dialogues, they often leave the reader uncertain about whether a statement formulated in the discussion really is to be taken as Plato's personal view. Yet on the subject of the doctor's task and competence, his texts are astonishingly unequivocal. His opinions about the concept of health vary a little, but keep a common core as they consistently look upon health as a kind of harmony, balance, and order. Much more difficult is to attribute to Plato one definite opinion about pleasure and its connection to good life.

When quoting Plato in English, I throughout use the translations from *Complete Works*, edited by John Cooper (1998). When commenting on or emphasizing a special Greek word or formulation, my sources are always the Greek texts presented in the Loeb Classical Library.

2. Definitions of Health

Plato is eager to emphasize that one definite concept of health exists, even if we may formulate different requirements for when a man or a woman, a youth or an elderly person is healthy (*Menon*, 72d–e). Plato generally defines health (*hygieia)* as a harmony, a balance, an order (*taxis, kosmos*), and disease (*nosos*) as a lack of balance, a disproportion (*ametria*), or a disorder. However, when we ask, "Between what entities is this balance or disproportion established?" he gives us different answers.

When Plato talks about health, he most often refers to bodily health. Some of his definitions explicitly mention components of the body and cannot be used in a discussion of pure mental health. However, in some passages, he is really talking about the health of the soul, which he identifies with virtue. Mental disease (illness), then, is a vice, a lack of balance or a disproportion, described by Plato as either licentiousness or ignorance (*Sophist*, 228d–e, *Timaeus*, 86b, *Gorgias*, 505b). In a few passages, Plato also considers health something that belongs to body and soul taken together as a whole.

Several formulations in the dialogues contain real definitions of health. I will here introduce five separate definitions.

A. Health consists in a harmony between parts of the body.

In the *Republic,* Plato says: "To produce health is to establish the components of the body in a natural relation [*kata phusin*] of control and being controlled, one by another, while to produce disease is to establish a relation of ruling and being ruled contrary to nature [*para phusin*]" (444d). Plato does not tell us what parts of the body he has in mind or what kind of control or ruling between the parts he is talking about. Perhaps he intends a division in head, breast, and trunk that corresponds to the three parts of the soul. In such a case, the mutual control that Plato speaks of may be interpreted as a kind of cooperation. When the body is healthy, the parts of the body cooperate well, and vice versa.

In the *Gorgias,* where Plato says that the regularity or organization of the body is healthiness (504c), we find a similar opinion.

Definition 1 gave us some problems of interpretation when we asked what parts or what kind of control Plato intends. While the translation above uses the phrase, "components of the body," H. D. P. Lee's translation uses "constituents of the body," and some translations into Swedish—for example, those by Claes Lindskog and Magnus Dalsjö—use the more concrete "parts of the body." But if we go to the Greek text, we find that Plato omits the noun: He talks about a relation between "those that belong to the body" (*ta en tō sōmati*). But the things he refers to can naturally also be things inside the body, not necessarily outer parts of the body, or they may also be bodily abilities.

A definition that is based on what happens inside the body is the following:

B. Health consists in a harmony between different kinds of stuff within the body.

This characterization can be found only in the *Timaeus* and is connected with Plato's strange "atomic theory," in which he combines the Empedoclean doctrine of the four elements with an idea that matter is composed of atoms with special regular shapes (regular bodies with faces built up of "elementary triangles"). Plato is mainly talking about a harmony or a proportion in the amounts of these four elements within the body. But he also means that the location of the stuff in the body is relevant for a harmonious and natural combination.

Plato says: "How diseases originate is, I take it, obvious to all. Given that four kinds of stuff exist out of which the body has been constructed—earth, fire, water, and air—it may happen that some of these unnaturally increase themselves at expense of the others. Or they may switch regions, each leaving its own and moving into another's region.—When these things happen, they bring on conflicts and diseases" (*Timaeus*, 82a).

In his further argumentation in the *Timaeus,* Plato describes motions of body fluids like blood, phlegm, and bile. Thus, he links up with the medical speculations of Hippocrates.

Plato introduces a third definition based on bodily components in the *Philebus* and the *Symposium*:

C. Health consists in a harmony between designated pairs of opposites within the body.

Plato gives us some examples of such pairs of opposites: warm–cold, wet–dry, more–less, rapid–slow, bigger–smaller. However, we must ask in what way pairs like these exist within the body, and how someone, as Plato says, "puts an end to the conflicts there are among opposites, making them commensurate and harmonious." I suppose he means that our body is healthy when, for example, its temperature is in some intermediate position between (too much) warmth and (too much) cold. But if so, how is it that the opposites still exist in the body when we have reached this intermediate position? I would say that they have now disappeared and are replaced by the intermediate position. However, Plato says that they exist, but are now in harmony. Socrates asks: "Is it not true that in sickness the right combination [*orthē koinōnia*] of the opposites establishes the state of health?" (*Philebus*, 25e). The answer is affirmative.

The same concept occurs in the *Symposium*: "The physician's task is to effect a reconciliation and establish mutual love between the most basic bodily elements. They are those that are most opposed to one another, as hot is to cold, bitter to sweet, wet to dry, cases like those. In fact our ancestor Asclepius first established medicine as a profession when he learned how to produce concord and love between such opposites" (186d–e).

The following definition takes as basic both bodily and mental elements:

D. Health consists in a harmony between the abilities of body and soul.

In the *Timaeus*, he says: "In determining health and disease or virtue or vice no proportion or lack of it is more important than that between body and soul" (87d). Plato aims here at a proportion (*xummetria*) in power between the mental and bodily abilities. Strength in the one must correspond to strength in the other; otherwise, the combination, the whole, will lack "beauty" (*ou kalon holon to zōon)* and result in disease. For "that combination of soul and body which we call a living thing" (87e) to be healthy requires a balance between the abilities. "When within it there is a soul more powerful than the body and this soul gets excited, it churns the whole being and fills it from inside with diseases" (87e–88a). Therefore, Plato prescribes: "The mathematician, then, or the ardent devotee of any other intellectual discipline should also provide for his body by taking part in gymnastics," and someone who has concentrated on developing bodily abilities should "apply himself to the arts and to every pursuit of wisdom" (88c). A problem with definition 4 is that it forces us to accept that also a harmony between very low levels in both mental and bodily abilities is healthy. I doubt whether that was Plato's intention.

Plato uses still another definition of health that emphasizes man as a combination or compound of body and soul. This definition does not restrict harmony to the strength of bodily and mental abilities. Instead, it presupposes harmony between the parts of the soul and of the body, and a harmony between the soul and the body that consists in the dominion of the soul over the body. The soul directs and controls the body. When this complex harmony and total balance occurs, then man is healthy. We must establish the dominion of soul over body and cure the soul from its disproportions. If we can warrant that kind of all-embracing structure, then a harmony in the details will appear automatically, in Plato's opinion. I will call this concept of health his holistic concept. I think we may compare these speculations to Plato's idea that the all-embracing structure of the just (and ideal) state with philosophers as a ruling class will establish the virtue of justice among the citizens. The justice of the state and the just moral character of its inhabitants will always appear together in Plato's opinion, but by ordering the state, the whole, in a just way, we create just people, not the other way around.

E. Health consists in a total harmony between soul and body and between the parts of the soul themselves and the parts of the body themselves.

This concept of health pervades the *Charmides*. Socrates says: "... just as one should not attempt to cure the eyes apart from the head, nor the head apart from the body, so one should not attempt to cure the body apart from the soul. And this, he says, is the reason why most diseases are beyond the Greek doctors, that they do not pay attention to the whole as they ought to do, since if the whole is not in good condition, it is impossible that the parts should be. Because, he said, the soul is the source of bodily health and bodily disease for the whole man, and these flow from the soul in the same way that the eyes are affected by the head. So we must first and foremost cure the soul if the parts of the head and of the rest of the body are to be healthy" (*Charmides*, 156e–157a). We must create temperance (*sōphrosunē*) in the patient's soul, for "when the soul acquires and possesses temperance, it is easy to provide health both for the head and for the rest of the body" (157a–b).

Evidently, Plato in these different texts is working with different concepts of health, a fact that causes problems when we wish to put his statements together into one theory. The definitions still have a similar structure: the idea of balance and harmony. However, many concepts used in the definitions are ambiguous. How should we interpret "components," "parts," "control," "harmony," and "balance" without the definitions being counter-intuitive or absurd? Plato also fails to provide a plain statement of his idea of the relation between body and soul, which is so central in, for example, definition 5. How are changes in the balance of the soul related to changes of the balance of the body? Are they sufficient or necessary, or sufficient and necessary for changes in the body?

Theories that take balance or harmony as something basic raise many questions. How do we know, for example, when balance is obtained? Can more than one state that means balance for a person exist at one time? Is balance always the same and the same for different persons, or does it vary? Plato's text gives answers to some questions of this kind. According to Plato, balance comes from the nature (*phusis*) of things. All things have their natural function, their natural location, their natural connections, etc. (see the quotations from the *Republic* and the *Timaeus* at definitions 1 and 2 above.) The doctors can, by a kind of rational investigation, find out where this natural balance is or ought to be in a single case. So, at a minimum, they know whether a body is healthy. The patient's opinion, however, is not so trustworthy, in Plato's eyes.

Also, what is natural is problematic. A natural balance is something good, something that we ought to strive for, if we have lost it. Plato's idea in definition 3, that balance is a concord of opposites and an intermediate position between too much and too little, bears something in common with Aristotle's thoughts of a golden mean that we should reach for. The doctor's rational investigation, which tells where the balance is to be found, also results in a normative insight. What is natural or according to the nature of things is at the same time, for Plato and other Greek philosophers, something descriptive and something normative, something that ought to be.

3. Health and Disease

Plato presupposes a close connection between man's bodily and mental functions. Even if the relation is far from specified, he evidently starts from the point that the soul and its properties determine the function and properties of the body. We should cure the soul in order to achieve balance also in the body. "It seems to me that a fit body doesn't by its own virtue make the soul good, but instead the opposite is true—a good soul by its own virtue makes the body as good as possible" (*Republic*, 403d; see *Charmides*, 156b–157c).

However, in the *Timaeus*, Plato appears at first glance to contradict this thesis when he says that some mental diseases are really caused by undesirable bodily states. We cannot therefore blame the subjects or hold them responsible for their actions, he concludes (86b–e). But according to Plato, these bad bodily states in turn are caused by the uneducated and unrestrained mental states of the young child. The parents and the surrounding society have permitted the child's mind to develop into and remain in such a condition. The parents and society are therefore responsible for the bodily state that predetermines the grown-up person to mental disease (87b). So even in a case like this, the properties of the soul, in a way, are Plato's starting point.

Aside from mental diseases, Plato's examples of diseases are plague and epidemic diseases, which are mentioned in a few passages. Most frequently, however, he speaks of diseases that are connected with our way of living our life. We might call them "lifestyle diseases." Plato occupies himself so extensively with these diseases that his texts give the impression that all diseases are such lifestyle diseases. Again and again he emphasizes that excessive and intemperate use of food and drink or a depraved and disorderly life cause diseases (see *Gorgias*, 517c–518e, *Republic*, 404a–405e, *Laws*, 691c–d). Both body and soul can also get into disharmony if not being tried and exercised, while training (*meletē*) and motion (*kinēsis*) are salubrious (*Theaetetus*, 153b–c).

Aging is not a disease, in Plato's view. When we are growing old, the elementary triangles in the substances of our bodies are getting worn out:

"the roots of the triangles are slackened as a result of numerous conflicts they have waged..." (*Timaeus*, 81c). The process ends in the dissolution of the body and the release of the soul. Death is the separation of body and soul. But when death is the result of a natural aging process, it is not painful: "The soul is then released in a natural way, and finds it pleasant to take its flight," but the death caused by "disease or injury is painful and forced" (81d–e).

According to Plato, health and disease are contradictory concepts, opposed to one another. You cannot at the same time (and in the same respect) be both healthy and ill, and if you stop being healthy you get ill, and vice versa (*Gorgias*, 495e–496a). In the *Timaeus*, when discussing how health and disease originate, he says: "And when this is the way it actually happens in each case, health will generally result. Disease, however, will result if things happen the other way around [*nosoi de, hotan enantiōs*]" (p. 82e). The contradictory character is also easily recognized in the formulations of the definitions. For example, according to definition 1, health follows nature and disease goes against or does not follow nature.

Ordinary, traditional medicine today also regards health and disease as two contradictory concepts. Disease is characterized as, for example, a condition of bodily organs, a kind of change in organs or a deviation from what is statistically normal. Health, then, is lack of disease. But traditional medicine does not normally acknowledge theories of health based on a concept of balance or harmony, as Plato's theory does. Exceptions are theories of physiological homeostasis—for example, Walter Cannon's from 1932.

However, today new health theories again propose that we should look on health as a kind of balance or equilibrium. The theories of Lennart Nordenfelt and Ingmar Pörn are especially relevant in the Scandinavian discussion of this topic. According to Nordenfelt, generally, a person is in health if and only if she has ability to realize her vital goals (Nordenfelt, 1987, pp. 76–80; see also Nordenfelt, 1993, pp. 94–101). Thus, when a person is healthy, a balance or equilibrium exists between what a person can do and what she in a deeper sense wants to do.

In such a theory of health—Nordenfelt calls his theory "holistic"—health and disease are not contradictory opposites. When a person cannot reach her vital goals, the person is unhealthy, but that does not necessarily mean that the person has a disease. Many causes for why a person cannot reach her vital goals are possible, from exterior circumstances to bodily capacities and mental factors. To Nordenfelt, a disease is a special kind of cause for ill health, not something that should be identified with ill health (see Nordenfelt, 1987, Chapter 4).

As to these two factors of comparison, Plato's theory takes an intermediate position between theories of traditional medicine and modern holistic equilibrium theories. A matrix can illustrate this comparison:

	The concept of health is:	The concepts of health and disease are:
Traditional medicine	not an equilibrium concept	contradictory concepts
Plato	an equilibrium concept	contradictory concepts
Modern holistic equilibrium theories (Lennart Nordenfelt)	an equilibrium concept	not contradictory concepts

4. The Doctor's Task

The task of doctors is to cure patients of disease, to create health. Chiefly, their mission is to reconstruct lost health (for example, *Charmides*, 165c–d, *Protagoras*, 345a, *Gorgias*, 452a, 478a). But the doctor should not take any preventive steps. Prevention, in Plato's opinion, is not a part of medicine. Instead, prevention is the task of the physical trainer, who makes our already healthy bodies stronger and more resistant against diseases. According to Plato, physical training also has a strengthening effect on the soul (*Republic*, 410b). But the art of the doctor is something useful only to the ill person (*Lysis*, 217a, *Republic*, 332e); if people were not ill, medicine would be of no value.

In Plato's opinion, gymnastics is superior to medicine. Gymnastics is more admirable, a finer thing (*kallion*), he says (*Gorgias*, 520b). I think Plato intends both that gymnastics is of more value to society than medicine, and that gymnastics is more fundamental. The physical trainer lays down the rules, like a legislator, and only when you break them does the doctor have to come in to restore balance (see the analogies in the *Gorgias*, 464b–465d). If people really followed the directives of their physical trainers, little need for doctors in society would exist (*Republic*, 410a–b). But people do not follow these directives. Therefore, medicine really accomplishes something good and is also needed in society (*Republic*, p. 341e).

Plato emphasizes that the doctor is an expert. If we want to be cured of our disease, we have to go to a doctor; no one else is expert in this field (*Crito*, 47a–c, *Laches*, 198d, *Republic*, 349e–350a; see also *Phaedrus*, 268b–c). We are not allowed to call in question a doctor's diagnosis and ordination. "Suppose the ordinary man thinks he is going to take a fever, and that his temperature will go up to a fever point; while another man, this time a doctor, thinks the opposite. Do we hold that the future will confirm either the one judgment or the other?" (*Theaetetus*, 178c). To Plato, it would be absurd not to trust the doctor's judgment (see *Lysis*, 210a).

In Plato's opinion, medicine (*iatrikos*) is an art or craft *(technē*). In the *Gorgias*, he explains what he means by "art" when talking about professions.

To practice an art is: (1) to do something in order to produce something good—here, good for the patient; and (2) to do it from an insight into the matter or subject—here, a knowledge of the nature of diseases and their causes. A practice that does not fulfill condition 2 but is based on experience and how things tend to be and behave Plato calls a "knack" or a "habitude" (*empeiria*). When a practice does not fulfill condition 1, and the aim instead is to please other persons, to give them what they want without thinking of what really is good for them, then the practice is flattery (*kolakeia*). Often those who carry on flattery give an air of practicing an art, and ordinary people also look upon those practitioners as doing that—which makes them and their activities even more objectionable, in Plato's eyes.

One example of flattery is pastry baking. "So, pastry baking, as I say, is the flattery that wears the mask of medicine." And baking is not an art or a craft but a knack, "because it has no account of the nature of whatever things it applies by which it applies them, so that it is unable to state the cause of each thing and I refuse to call anything that lacks such an account a craft" (465a–b). More strictly, Plato says that he refuses to call a practice a craft if the practice is irrational—that is, not based on intellectual insight (*egō de technēn ou kalō, ho an ē alogon pragma*). But medicine is an art. "I said that ... medicine has investigated both the nature of the object it serves and the cause of the things it does, and is able to give an account of these" (501a; see *Phaedrus*, 270e).

In the *Philebus*, Plato does not draw such a sharp line between what is an art and what is not. Instead he says that different arts are compounds of two parts, one exact and intellectual, the other inexact and empirical. They are in various degrees allied to knowledge (*epistēmē*) and connected to concrete practicing. The arts, according to Plato, can be more or less pure. The arts "in the hands of the philosophers" have more precision and are more pure than other arts (57c). But medicine is a real mixture of both components (56a–b) and is placed somewhere in the middle on the purity scale of arts.

Even if medicine is founded on an intellectual insight, to be a good doctor you need practice and experience, according to Plato. In the *Republic*, Socrates states: "The cleverest doctors are those who, in addition to learning their craft, have had contact with the greatest number of very sick bodies from childhood on, have themselves experienced every illness, and aren't very healthy by nature" (408d–e).

Doctors are experts, we have learned, and experts never fail (*Republic*, 340d–e). Still Plato sometimes says that bad doctors exist. A bad doctor does not succeed in curing, but instead makes the conditions even worse for the patient (*Protagoras*, 340d). However, while even a good doctor may make mistakes, "if he happens to slip, he can put things right" (*Republic*, 361a). To be a doctor, you would have to fulfill requirements 1 and 2 above. You can fulfill 1, having an intention to do something good for the patient, without really succeeding in doing something good. I think Plato believes that if you fulfill 2, having a real insight in medical matters, then things connected to the patient's condition will happen just as you believe they will happen. Therefore, if you are a doctor and thus meet conditions 1 and 2, then you will, except in extraordinary circumstances, also be a good doctor. I think that is why Plato is so sure that doctors are experts. If you do not fulfill 2, then you are no doctor at all, and, consequently, you cannot be a bad doctor either (*Protagoras*, 345a–b).

To be a good doctor, you must know the rules and laws of the art of medicine. The fulfillment of condition 2 will warrant that. But the insight or understanding that is included in 2 will also include knowledge about when the rules should be applied and when they are not applicable. The good doctor is a good judge on such matters (*Statesman*, 295c–296c). Good doctors never act thoughtlessly simply from established rules. Their judgment tells them what they and their art can do and what they cannot manage. "A first-rate captain or doctor, for example, knows the difference between what his craft can and can't do" (*Republic*, 360e–361a).

The doctor's task is to cure patients of disease, which means re-establishing natural balance. Sometimes the curing is referred to as purification (*katharmos*) of the body only or of both body and soul. "They [the doctors] draw off the worst and leave the best" (*Republic*, 567c; see also *Sophist*, 226d–228e). The curing is without exception described as something painful and trying to the patient. Medicaments with disagreeable tastes, a plain and strict diet, and the doctor's carving and burning the patient are examples. Yet Plato most frequently emphasizes the change of way of living, and the abstaining from different pleasures—sometimes from all kinds of pleasures—because in his opinion, the enjoying of unworthy pleasures is the main cause for diseases. However, the patient chooses this painful treatment consciously and freely in order to regain health—that is, a greater good (see, for example, *Gorgias*, 467c, 478b–c, 505a–c, *Protagoras*, 353c–354b, *Statesman*, 293b, *Laws*, 646b–c).

In two passages in the *Laws,* Plato mentions that a free doctor (in this dialogue Plato also assumes the existence of slave doctors) explains to the (free) patient the nature and the cause of the disease. When the patient understands what has happened, the doctor tells what he has to do to cure the patient from the disease, and then the patient agrees to that course of treatment (*Laws*, 720d). "This doctor would be acting almost as a philosopher, engaging in a discussion that ranged over the source of the disease and pushed the inquiry back into the whole nature of the body" (857d). Even if Plato's statements here appear to be descriptive, their contents must be normative. The free doctor should teach and educate the patient in this way. What Plato says may even suggest an early introduction of what we today call a principle of "informed consent."

But we must remember that the doctor in his eyes is a kind of artisan, someone you send for when you need skilled assistance. If you want to be helped, you have to accept what the doctor suggests. You also have to pay doctors—you buy their service—but they are the experts. I do not think Plato meant that the patient could question what the doctor proposes and suggest another kind of treatment; but the patient is free to decide not to buy the doctor's help. However, Plato's insisting on the doctor's informing and educating the patient is significant, and can be taken as an ethical principle for doctors.

Plato's texts contain other ethical principles for doctors. In the *Republic,* Plato emphasizes that even if the doctor is paid and in one way is a businessman, "no doctor, insofar as he is a doctor, seeks or orders what is advantageous to himself, but what is advantageous to his patient" (342d; see also 346a–c). Doctors should never intentionally damage a patient, neither for their benefit, nor at a request from the patient's relatives (*Statesman*, 298a–b). A doctor must not make a patient's condition worse by trying to satisfy the patient's wants, when these are debauched. "Everything sound and healthy in the body must be encouraged and gratified; that is precisely the object of medicine. Conversely, whatever is unhealthy and unsound must be frustrated and rebuffed; that's what it is to be an expert in medicine" (*Symposium*, 186c).

Plato meant also that the doctor in some cases should refuse to treat a patient. He says that Asclepius, demigod and inventor of medicine, reserved his art for people who were prepared to live ordinary and temperate lives. If you have destroyed your body so that you cannot resume temperate habits, then Asclepius would not help you. "Therefore, won't we say that Asclepius ... taught medicine for those whose bodies are healthy in their natures and habits but have some specific disease? His medicine is for these people with these habits. He cured them of their disease with drugs or surgery and then ordered them to live their usual life so as not to harm their city's affairs. But for those whose bodies were riddled with disease, he didn't attempt to prescribe a regimen, drawing off a little here and pouring in a little there, in order to make their life a prolonged misery and enable them to produce offspring in all probability like themselves. He didn't think that he should treat someone who couldn't live a normal life, since such a person would be of no profit either to himself or to the city" (*Republic*, 407c–e).

Thus, in the end the argument is a moral one. The doctor should not help such people to go on living, because they are of no good to the state. On the contrary, they are bad examples for others, Plato would say. And if we help them to go on living, they may even have a chance to give birth to individuals of the same shameless character. Before deciding about any treatment, the doctor has to investigate whether a treatment would make the state better. Only if it would improve the state should it be administered, in Plato's opinion—a demanding and, in our eyes, provocative ethical principle.

The doctor is an expert on health and how to cure a patient of diseases. Where on a social ranking would Plato place such a person? He offers a fairly detailed answer to that question in the *Phaedrus*. There Plato gives us a social ranking when he speculates about reincarnation and in what kind of living creature a soul of such and such a character will be born or reborn. "A soul that has seen the most" will be born as a philosopher, the best out of nine possibilities. The next best alternative is to be born as a lawful king or a competent commander. The worst is to be born as a tyrant, and second worst as a sophist or demagogue. Plato suggests that "the fourth [best] will be a trainer who loves exercise or a doctor who cures the body" (248d). Plato thus places the doctor somewhere in the middle of the list. Plato still respects doctors (and trainers) as experts, but they cannot in his eyes compete with, for example, philosophers, and lawful rulers.

Plato takes his list for granted, offering it without argumentation, but we can easily imagine what kind of arguments he might use. Five points of comparison between a doctors and philosophers that Plato might have suggested in order to justify his ranking are: (1) doctors are wage-earners, philosophers are not; (2) a doctor's art is closely connected to concrete actions, while the art of the philosopher is more loosely connected to actions; (3) the domain of the doctor is mainly the body, and the domain of the philosopher is the soul; (4) philosophers have insight in deeper, weightier questions than doctors; and (5) the art of the philosopher is pure, but the art of the doctor is mixed, partly empirical (see *Philebus*, 56a–57d).

Health, according to Plato, is not the most significant element of human life. We must not concentrate our interest and ambition on our health or on questions about health and disease. Man is healthy by nature, and he should carry out his share of work for the state. To keep alive at any price, without being able to take part in the work for the state, is not a good. On the contrary: When people concentrate on their health and want to consult a doctor anytime and about any question, that is a sign of unsound conditions in a society.

Neither should doctors be given power of that kind. We consider it "most shameful," according to Plato, when people leave the responsibility for their life to others. "But excessive care of the body, over and above physical training, is pretty well the biggest obstacle of all. It's troublesome in managing a household, in military service and even in a sedentary public office. Yet the most significant of all, surely, is that it makes any kind of learning, thought, or private meditation difficult ..." He ends his warning: "Hence, wherever this kind of virtue [that is, intellectual virtue] is practiced and examined, excessive care of the body hinders it, for it makes a person think that he's ill and be all the time concerned about his body" (*Republic*, 407b–c).

Modern people, citizens of the twenty-first century, take a great interest in their health—too great an interest, in Plato's eyes. As recent investigations have shown, people in Western societies reckon health as something of fundamental value, often as the fundamental value. Today many people look at doctors as authorities on general subjects and in areas far beyond the field of medicine, which gives the medical profession an especially high status. To Plato, these facts would spell out a situation that was strongly undesired and even dangerous. I think that Plato, if he knew about modern attitudes toward health and doctors, would be alarmed and sincerely concerned about the survival of Western world.

5. The Art of Caring: Doctor, Statesman, Judge, and Philosopher

According to Plato, doctors do not belong to the highest-ranked social groups, neither is their knowledge of utmost value. Yet they are a group in which Plato places great interest. The reason is, I think, that medicine is a caring (*epimelētikos*) art or expertise. In Plato's eyes, the duties of politicians and judges, too, are caring activities (see *Statesman*, 275e–276c)—even some respects of the philosopher's activities may be reckoned as caring. Therefore, Plato points out several analogies between, for instance, the aims of these arts, the practitioners' presupposed knowledge, and their methods and results. In understanding what a good doctor is, the reader easily understands what Plato intends a good judge or a good statesman, for example, to be.

Among these caring arts, Plato distinguishes between a ruling art and a serving, subordinate (*hypēretikos*) art, an art that is ruled by other arts. All curing arts—for example, medicine—are ruled arts. And the highest ruling art, the "kingly" (*basilikos*) art, is politics, diplomacy (*Statesman*, 305a–e). Evidently, Plato considers the ruling arts as primary, because they establish goals and formulate rules for how people can reach these goals. The curing arts are secondary, as their aims derive from the ruling arts, and because their task is to restore balance when people do not follow the rules established by the ruling arts. If people really followed these rules, no curing arts would be needed (see, for example, *Republic*, 410a–b, *Laws*, 628c–d).

The caring arts presuppose a special kind of knowledge. Practitioners need an insight into what is best for their clients, whether they are patients, criminals, or states, or whether you are working with the body, the soul, or with classes in a society. This optimum is always a kind of balance or harmony. As a good political leader, judge, or doctor, you have to know the right proportions and the right measures (see *Statesman*, 284a). In some passages, Plato stresses how rare people possessing such competencies are. In his opinion, but few "real" statesmen, "real" doctors, and "real" judges exist (*Statesman*, 292d–293e, *Gorgias*, 521d). Most people working in these fields instead practice flattery. They try to give people what they want or what they say that they want, and they have no insight in what is best for people (see, for example, *Gorgias*, 511–519).

The curing treatment, sometimes characterized as a purgation, is always described as something painful or disagreeable. The doctor's and the judge's work are often compared to one another (*Gorgias*, 478a–479c, *Sophist*,

228e–229b). Their clients have to go through a painful treatment in order to restore balance in body or soul; that is, to get the best for themselves. About punishment Plato says: "Now, wasn't paying what's due getting rid of the worst thing there is, corruption?—It was.—Yes, because such justice makes people self-controlled, I take it, and more just. It proves to be a treatment against corruption.—Yes" (*Gorgias*, 478d).

Not every balance can or should be restored, in Plato's opinion. The doctor should not try to cure patients who are unwilling or unable to live normal, temperate lives. Nor should all criminals get a punishment that will make them atone for their crimes, restore balance in their souls, and make them fit to return to society and live ordinary lives. In some cases, Plato explicitly defends capital punishment (*Republic*, 409e–410a, *Laws*, 735d–e; see also *Gorgias*, 512b: "He [a helmsman; here, a man with insight] knows that for a corrupt person it's better not to be alive, for he necessarily lives badly").

A more in-depth elucidation of Plato's ideas of the relations between these professions within the art of caring and an analysis of what he says about differences and identities in aims, knowledge, methods, and result would be an interesting task. However, it would extend beyond our present purpose. So, what I emphasize in this article is solely their overall kinship.

The kinship is also evident through Plato's choice of words. He often uses medical terminology when he talks about political or moral subjects. A society or a state can, in his words, be healthy or sick (for example, *Republic*, 556e, 563e); can be in a condition of fever; or can cure a state of a disease, etc. Also, the philosopher educates and cures the client from ignorance and vice, which are diseases of the soul (see *Republic*, 444d, *Sophist*, 228e–229c). The parallelism between the medical concepts of health and disease and the moral concepts of virtue and vice is consistently carried through in the dialogues. Thus, the ideas of Plato, or perhaps even Socrates, must be taken as a starting point for the parallelism between philosophy and medicine that later became explicit in Aristotle and in Hellenistic authors (see Nussbaum, 1994).

In the *Charmides,* Socrates is jokingly introduced as a doctor who can cure headache (155b). But when the discussion starts, we find that that description was not entirely a joke. What Socrates does can easily be—and is—described in medical terms (see the converse comparison in the *Laws*, 857d, where the doctor "would be acting almost like a philosopher"). In the *Gorgias,* Socrates also declares that he himself is a true politician, the only one living in Athens. "I believe that I'm one of a few Athenians—so as not to say I'm the only one; but the only one among our contemporaries—to take up the true political craft and practice the true politics" (521d). Socrates' statement should not be taken in jest, even if it at may first strike us as surprising and inexplicable.

But in Plato's opinion, the philosopher is the one who has the deepest insight, irrespective of the subject matter. The philosopher has beheld the world of ideas, and therefore has an insight into the aims and functions of everything existing. Thus, he is most deeply acquainted with the aims and functions of the state and, therefore, also the best statesman. Philosophers must be kings or kings must be philosophers, the famous Platonic credo says (*Letter VII*, 326b; see also *Republic*, for example, 375e, 376c and Book VI). Therefore, philosophy (or a part of it) belongs to the art of caring, as a ruling and as a curing art. In Plato's opinion, philosophers can be statesmen, but they are also a kind of doctors.

6. Pleasure and the Good Life

How should we live our lives? Plato's answer is short, but not simple: We should always let our reason, our intellectual faculties, guide and control the other faculties. "[T]he self-controlled man, because he's just and brave and pious, as we've recounted, is a completely good man," and therefore "blessed and happy"(*Gorgias*, 507c). I think that Plato would describe such a person as healthy. To have insight, to be virtuous and just, and to be happy: In Plato's opinion, these things go together; you cannot have only one of them.

This is the essence of a good life, a harmonious life. The biggest threat against a good life and against virtue and health is our uncontrolled desire for pleasure. Is pleasure, then, in itself something bad? In some passages, Plato explicitly adds pleasure to the list of components of the good life. The good, self-controlled life is more pleasant than all alternatives (see *Laws*, 734d–e).

While Plato makes comments on pleasure in most of his dialogues, we cannot easily get a comprehensive overview of his statements or to bring them together into one theory of pleasure. I will not try to do that. But as his opinions about pleasure are relevant for what he says about health, the doctor's task, and what a good life can be, I will make some remarks about what he says.

Generally, Plato must be characterized as a non-hedonist. He will not identify what is good with what is pleasant, and he argues vigorously that the "good" and "the pleasant" or "pleasure" are different things. He would strongly deny that all pleasant things are good; some pleasures are definitely bad, in his view.

In the *Protagoras,* a part of the discussion indeed presupposes an identification of the good with pleasure. But Plato does not affirm or argue for this thesis in this dialogue or in any other text in the *Corpus Platonicum*. Socrates in the *Protagoras* instead shows that even if we would accept this identification, as he believes the average man does, we would still have to allow a place for insight and rational calculation when deciding what things or actions are good. (I align here with the interpretation given by A. E. Taylor,

1960, p. 260.) Therefore, the discussion in the *Protagoras* implies no objection to my general characterization above.

In one dialogue, the *Phaedo*, Plato is totally negative about pleasures. He suggests that all pleasures are bad and should be avoided. The central thing for a person, or for a philosopher, is to "acquir[e] knowledge," let "the soul grasp the truth," and, "in reasoning," let "any reality" become "clear to the soul" (65a–c). But anything that is connected with the body disturbs this work of the soul: "And indeed the soul reasons best when none of these senses troubles it, neither hearing nor sight, nor pain nor pleasure, but when it is most by itself" (65c). Therefore, in Plato's view, a philosopher must welcome death and even work for his death, the separation of body and soul. The negative value of pleasure here is explicitly connected to its effects. The reason is that experiencing pleasure disturbs our thinking about "realities" that pleasure is bad.

In the *Phaedo,* Plato identifies pleasures with some pure bodily experiences. Pleasures nail the soul to the body. "Because every pleasure and every pain provides, as it were, another nail to rivet the soul to the body and weld them the together. It makes the soul corporeal, so that it believes that truth is what the body says it is" (83d). In other dialogues, Plato makes distinctions between different kinds of pleasures, but not in the *Phaedo*. Here all pleasures must be taken as bodily pleasures; that fact can also explain his totally negative attitude toward pleasures.

In many dialogues, Plato reckons some pleasures as good and some as bad. However, his valuation is not based on exactly the same distinction everywhere. I will briefly present his different distinctions.

(1) In the *Gorgias,* Plato separates pleasures based on the value of their effects. "The thing you are saying now, evidently, is that some pleasures are good while others are bad. Is that right?—Yes.—Are the good ones the beneficial ones, and the bad ones the harmful ones?—Yes, that's right.—And the beneficial ones are the ones that produce something good while the bad ones are those that produce something bad?—That's my view" (499c–d). Plato then specifies what these good or bad effects are. Health and strength are then suggested as something good, while disease is something bad.

In the *Protagoras*, we find a distinction of the same kind (351b–c, 354a–d). But here, as Plato is arguing from a kind of hedonistic point of view, the effects are limited to future pleasures and pains; the pleasure that produces a better balance of pleasure over pain is a higher pleasure. We have here embryos of the concepts of fecundity and purity in Bentham's moral philosophy and a forerunner to his argumentation: "sum up" and "take the balance" (Bentham 1996, pp. 39–40.)

(2) The *Philebus* hints at another distinction. Here, Plato suggests that pleasures that are pure [*hēdonas... katharas*]—that is, not mixed with any kind of pain—are better than the impure pleasures (66c).

(3) In the *Philebus,* Plato also connects the value of a pleasure with the truth or falsity of the belief to which it is attached (37e–41e). Plato does not say that every pleasure is attached to a belief, but when a pleasure is attached to a belief—for example, when I am glad because I believe that something is going to happen or has happened—then my pleasure is a better one if my belief is true, and a bad one if it isn't. Perhaps we can refer to pleasures as being justified or unjustified. Plato even uses the concepts of truth and falsity: "Nor, I presume, will we find any other account for badness in the case of pleasures unless they are false" (40e).

(4) However, the most central of his distinctions is his separation of bodily and mental pleasures. It is implicit in the texts, present in several passages (for example, *Philebus,* 12c–d). In the *Republic,* Plato is explicit on this point. In his opinion, the soul is divided into three parts with separate and different desires. "The first, we say, is the part with which a person learns, and the second the part with which he gets angry." The third part is "called the appetitive part, because of the intensity of its appetites for food, drink, sex, and all the things associated with them, but we also called it the money-loving part, because such appetites are most easily satisfied by means of money" (580d–e). As the parts have different desires, different pleasures correspond to them. Thus, three kinds of pleasures exist.

Plato also supposes that three kinds of people exist, if we look at the part of the soul that rules the person's life. People are philosophic, victory-loving, or profit-loving. But which pleasure is the best one? We must decide that, as he says, "by experience, reason and argument" (582a). Whom shall we ask? To Plato, the philosopher is the only possible candidate. A philosopher can reason and argue, but is also the only one acquainted with all the three kinds of pleasures. "A philosopher has of necessity tasted the other pleasures since childhood, but it isn't necessary for a profit-lover to taste or experience the pleasure of learning the nature of things that are and how sweet it is" (582b). Profit-lovers cannot do so even if they try, Plato says, nor can victory-loving people. Thus, the only competent judge in this question is the philosopher (582c–e), and the conclusion is evident: "Then, of the three pleasures, the most pleasant is that of the part of the soul with which we learn and the one in whom that part rules has the most pleasant life" (582e). The other pleasures are not so pleasant and of less value.

John Stuart Mill's famous discussion in *Utilitarianism* about the qualities of different kinds of pleasures in many details corresponds to what Plato says in the *Republic*. Mill uses exactly the same strategy, asking: Whom shall we ask? His requirements of competent judges are exactly Plato's: intelligence and acquaintance with all kinds of pleasures (Mill, 1972, pp. 8–11).

In most passages, Plato uses this distinction in a simpler form: He distinguishes just between pleasures of the soul and pleasures of the body. For example: "Then, when someone's desires flow towards learning and everything of that sort, he'd be concerned, I suppose, with the pleasures of the soul itself by itself, and he'd abandon those pleasures that come through the body" (*Republic*, 485d). His idea in the *Republic* is that the mental pleasures are good and contribute to the value of, for example, the life of the philosopher. The bodily pleasures are of less value. Some of them are, in Plato's eyes, definitely bad and shameful.

The *Republic* does not give us any exact answer how we should treat the bodily pleasures. We can, however, use distinction 1 above, from the *Gorgias*, where Plato talks about the effects of pleasures. The quoted text (499c–d) refers mainly to bodily pleasures. If we look at these two texts together, we may say that Plato asserts that (1) mental pleasures are generally good, and (2) bodily pleasures are good or bad depending on their effects. Plato would assume that (too) great amounts of bodily pleasure always cause bad conditions—for example, diseases—and therefore are bad.

If we are to connect what Plato says in the *Phaedo* to this interpretation, we must suppose, as I did above, that when he talks about pleasure in that text, he refers solely to bodily pleasures, leaving out the mental pleasures. His statements about pleasures in the *Phaedo*—for example, their direct contact with the body, and their "opposition" to our mental faculties—also fits well into such an interpretation. But his position there remains extreme.

Is pleasure something we can do without? In the *Phaedo*, the answer must be "Yes." The philosopher should do without pleasure. Nor are the mental pleasures, for example, in the *Republic*, emphasized as something necessary for man's life and really worth aiming at. Only in a few passages in the *Corpus* does Plato hold out pleasure as something relevant or a necessary component in a good life (for example, *Laws*, 734c–e). In the *Philebus*, he explicitly tries to prove such a thesis.

Socrates tells us that he had a dream that neither knowledge nor pleasure is "the good"; the good is something different from both. In order to prove that, Socrates asks us to compare two lives: a life of knowledge and a life of pleasure. "Let there be neither any knowledge in the life of pleasure, nor any pleasure in that of knowledge. For if either of the two is the good, then it must have no need of anything in addition. But if one or the other should turn out to be lacking anything, then this can definitely no longer be the real good we are looking for" (20e–21a).

Socrates then asks a young man, Protarchus, whether he would find it acceptable to spend his whole life enjoying the greatest pleasures. The answer is an unequivocal "Yes." He continues: "And would you see yourself in need of anything else if you had secured this altogether?" Protarchus first answers "No." But when Socrates points out that he cannot presuppose that he should be in possession of reason, memory, and true opinions, Protarchus regrets his answer. Socrates argues: "But, not possessing right judgment, you would not realize that you are enjoying yourself even while you do, and, being unable to calculate, you could not figure out any future pleasures for yourself. You would thus not live a human life but the life of a mollusk or of one of those creatures in shells that live in the sea" (21b–c). But this is a life not worth choosing. Conclusions: Pleasure alone is not enough, and cannot be the good, and knowledge is necessary for a good life.

Then Socrates repeats the argumentation and asks if the young man instead would choose a life in possession of "every kind of intelligence, reason, knowledge, and memory of all things, while having no part, neither large nor small, of pleasure or of pain, living in total insensitivity of anything of that kind" (21d–e). But Protarchus cannot accept such a life either. Conclusions: Knowledge alone is not enough, and cannot be the good, and pleasure is needed to make the life a good life.

Two other conclusions to be drawn are, firstly, that, as in Socrates' dream, the good must be a third element, in addition to knowledge and pleasure, and secondly, that knowledge and pleasure are both necessary parts of a good life. The best life must therefore be a mixture of knowledge and pleasure. "Everybody would certainly prefer this life to either of the other two, without exception" (22a; see also 27d).

The structure of the first part of the argument is the same as that of two well-known arguments in modern moral philosophy, both aiming at the conclusion that "the good" cannot be identified with pleasure or pure satisfaction: J. J. C. Smart's arguments about the man with the electrodes (Smart and Williams, 1973, pp. 18–22) and Robert Nozick's experience machine (Nozick, 1974, pp. 42–45).

We cannot easily to bring Plato's statements about pleasure together into one theory. Plato in the *Philebus* explicitly chooses not to divide pleasures into different classes when he argues about the necessity of pleasures in the good life. He says that he has no use for such a distinction here, even if it could be made (20c). Therefore we cannot, as we might have wished, draw a conclusion that some special kinds of pleasures are necessary, while other pleasures are not.

7. Final Remarks

Pleasures and uncontrolled desires are the main causes for diseases, according to Plato. People live licentious lives that destroy their health. Therefore, after an informative presentation of the diagnosis, and, possibly, after treatments such as carving and burning and the patient's partaking of some unsavory medicine, the doctor's prescription must be an order to live a temperate life or even to abstain from some kinds of pleasures.

As we saw, some pleasures are still necessary for the patient to live a good life. As Plato deems mental pleasures as thoroughly good, the pleasures he has in mind when he talks about causes for diseases and about pleasures that are consumed in too large proportions must be purely bodily. The doctor must find out the proper balance for each patient. To keep healthy, the patient has to follow the doctor's prescription and practice self-control, which, in Plato's eyes, is a virtue. Thus, the patient fell ill because of a vice, intemperance, and the doctor instructs him or her how to become virtuous. The parallelism between health and virtue and between the areas of medicine and morals is evident. The doctor's and the philosopher's prescriptions are identical: They both practice a caring art.

Two

HEALTH AS METAPHOR AND AS METONYMY—AN EXPLORATION

Per Sundström

1. Introduction

This essay is about "health" as metaphor and as metonymy, two figures of speech that depart from the literal sense, or senses, of "health." The essay presents examples of metaphorical and metonymical uses of "health," and demonstrates frequently occurring ambiguities regarding the interpretation of phrases incorporating "health." The essay also broaches the interplay between literal and figurative senses of "health."

This is a first exploration of a new field of linguistic-semantic investigation. No doubt it deserves further attention.

2. Metaphor and Metonymy Explained and Exemplified

A metaphor is a figure of speech in which a word that (literally) means one thing in one specific domain of reality and meaning is adopted into another domain, where it (metaphorically) will mean something distinctly different while retaining some resemblance to what it meant in its original domain. In other words, metaphor is letting a word that naturally belongs in one semantic context stand for a thing that belongs in another context.[1] A few examples will clarify:

"President Clinton is a fox; it has proven difficult to hunt him out." Here both "fox" and "hunt" are metaphors, words adopted from one domain into another, their meaning thereby being changed while retaining some resemblance to their original meaning.

"She is drop-dead beautiful." Here the compound adverb "drop-dead" is a metaphor, a word adopted from a context of dread and disaster into one of exaltation and pleasure.

"Madonna is one of the biggest stars ever in pop music." Here "star" is a metaphor, though currently almost "dead" (which is itself a metaphor) since we have got so used to it that we tend to overlook the transformation of meaning from the context of astronomy/astrology to that of popular culture. "Madonna" too may be regarded as a metaphor, an ironic metaphor—when the word is not taken to be simply the proper name of this particular American pop singer.

A metonym—or, as some prefer to say, an instance of metonymy—is a figure of speech which lets a word stand for a thing that is related to that which the word literally designates within a common domain of reality and meaning. The relatedness here referred to may be one of cause and effect, or part and whole, or closeness (and possibly others).[2] A few examples will illustrate the point.

"I love Plato, but Aristotle is a bore." Here "Plato" and "Aristotle" are metonyms, words standing for the writings authored by Plato and Aristotle, respectively. The names of the causes stand in for the effects, the producers for the products.

"Sweden beat England two nil last night in Wembley Stadium." Here "Sweden" and "England" are metonyms, words standing for the two soccer teams of Sweden and England, respectively. Wholes stand for parts, a national team being but a tiny part of a nation's population. Alternatively we might say that cause stands for effect, each nation having produced whatever skills and team spirit their national teams manage to muster.

"Unfortunately he has taken to the bottle recently." Here "bottle", which is a metonym, stands for that which is contained in, and may be poured out of, (some) bottles, namely alcoholic drink. The bottle is close to the drink and contains it; moreover, a bottle is part of the whole that is made up of a bottle and its contents (a bottle of wine, for example).

3. Lexical Definitions of Health

Useful, and not so useful, lexical definitions of "health" abound. Four of them, chosen at random, run thus: "the state of being well in body or mind,"[3] "soundness, especially of body or mind,"[4] "freedom from disease or abnormality,"[5] "a condition of optimal well-being."[6] The positive connotations here as regards value are unmistakable: health is either positively good, or else the laudable absence of things that are positively bad.

But expressions like "good health," "bad health," "poor health," and the like, make things more complicated semantically, even as they suggest that health, good or bad or perhaps neutral, is something you will have to live with as long as you live. These expressions correspond to definitions of health such as: "a person's mental or physical condition";[7] or "the overall condition of an organism at a given time."[8] Unlike the former set of definitions, these are value-neutral, scientifically-minded definitions, which invite further specification in terms of the various conditions which the trained, and technologically enhanced, senses of a medical professional may perceive. However, the value-neutral, scientific descriptions of the organism that will come out of this will typically not remain value-neutral as they are assembled to give a picture of "the overall condition of an organism at a given time." They will then be coupled with value predicates such as "excellent," "good," "bad," and "poor." Because no real, multi-dimensional human being will accept being looked upon as a set of value-neutral descriptions; and no doctor will get away for long with a value-free look upon his or her patients. Both parties will need phrases like "in good health" and "in poor health" as long as they make use of the word "health" at all.

We may conclude that sooner or later "health" will receive value predicates—or become positively value-laden as in definitions such as "the state of being well in body or mind."

A further semantic complication, on the lexical level of signification, is the use of the word "healthy" to signify not only, trivially, "possessing good health,"[9] "conducive to good health,"[10] and the like, but also things like "beneficial, helpful,"[11] and "likely to be for one's good."[12] These uses of the word draw on its positive value connotations, which can be applied to various things within many contexts other than that in which "healthy" originally belongs. In other words, "healthy" may serve as a useful metaphor, which can be used extensively: "a healthy afternoon"; "a healthy occupation"; "a healthy discipline"; "a healthy environment"; "a healthy atmosphere"; "a healthy friendship," etc. But as these examples show, "healthy" may also, and even simultaneously, serve as a metonym: a healthy afternoon might, literally, improve your health, especially if spent in places like Venice or Florence; and a healthy occupation may—in addition to being good for you (in general)—be conducive to health (of mind and body). And so on.

In other words, many phrases which incorporate "healthy" may be interpreted either in the direction of metaphor, or in that of metonymy. Sometimes the context in which such phrases are uttered or written indicate which interpretation is the preferable one. But that is not always the case. A mixture of both may be intended by the speaker or writer, and heard, felt and understood by the interlocutor or reader. Both are possible, and they need not be mutually exclusive. On the contrary, they may enhance each other.

A similar mixture of metaphor and metonomy—or possibly an oscillation between them—is seen in more complicated definitions of "healthy," such as: "indicative of sound, rational thinking or frame of mind."[13] We are here dealing with a metaphor insofar as a healthy mind is one that is characterized by intellectual virtues rather than by its robustness. But an admixture of metonomy is possible and plausible, and sometimes unmistakable: a healthy utterance may be one that issues from, and so points to, a healthy, robust mind.

4. The Metonymy of "Health Care"

When we speak about health care, it appears obvious that "health" means precisely that: health. It is health that we care about, and by engaging in practices subsumable under the heading "health care" we intend to support and promote health. This might seem plain enough: "health" in "health care" means, literally, health.

But on further reflection what we care for in health care is not obviously health in the literal sense. Most of our caring activities—including our attempts to cure and to mend—are not directed, literally, at health but at the body and the organism and its proper functioning. This is what we, directly and literally, care about—at least most of the time. By caring for the body and the organism we *affect* their health; we hope that what we do will *lead* to health. In other words, the relation between "health" in "health care" and health proper is a causal one; changes brought about in the body and organism bring about health, literally health. And so "health" in "health care" is a metonym instead of a signifier that literally stands for health, health proper, health itself.

Health itself may be regarded as an abstraction from the down-to-earth concerns of physicians and other professionals in health care, or as an imponderable state of being of your self, which is somehow a result of what happens in and with your body (including its mind, soul, or spirit). Either way health is something that is reached and affected only indirectly, by way of that to which you have direct access and which you can manipulate: namely, the body and the organism.

In other words, you reach the imponderable by way of the ponderable. In health care we care for the body and organism, which we can control and manipulate, in order to produce that elusive—yet for the individual, crucial—thing, health. This indirect, and causal, relation between "health" in "health care" and health proper turns the former into a metonym.

Metonymy, here as elsewhere, may be seen as a sort of shorthand that makes verbal communication quicker, and possibly more effective. But you must be an insider, who knows important things about what is being spoken of, in order to profit from metonymy. An outsider may be confused rather than enlightened by language that makes frequent use of metonymy. For example, he or she may confuse health and the human body and begin to believe that the ultimate aim of health care is intact, or mended, bodies—"health"—not health itself.

5. Metonymical Celebrations: "Healthy Houses" and "Healthy Cities"

A prevalent metonymical use of "healthy" is found in expressions such as "healthy houses" and "healthy cities." Though "healthy" might be a metaphor here, metonymical interpretations would be preferable in most contexts. A healthy house is first and foremost a house that advances, or at least does not have serious adverse effects on, the health of its inhabitants. Similarly, a healthy city is one that furthers the health of its inhabitants, typically by reducing environmental health hazards, such as air and water pollution, and noise and heavy traffic, which are bound to emerge in big, and growing, urban areas. In both instances, we are dealing with cause-effect metonymy.

Metaphorical interpretations of "healthy houses" and "healthy cities" are more far-fetched, but they are by no means mere ad hoc inventions. We may consider a house healthy if well-built, solid, and well-proportioned. The metaphor here draws on the imagined similarity between a human body that is well-built and solid, and to that extent healthy, and a house that is a well-built and solid construction. A similar idea of solidity, order and proper functioning lies behind a metaphorical interpretation of "healthy" that turns a healthy city into one that is endowed with wise, considerate leadership.

But again, metonymy is the rule, and metaphor the exception, when we encounter "healthy houses" and "healthy cities." Both healthy houses and healthy cities, as metonymically interpreted, have been made the objects of organized pursuit and celebration in recent years. I will give two examples of this.

In a booklet entitled *The Healthy Building* we find a sustained meditation on the concerns and processes of healthy building, of the erection of "healthy buildings."[14] Ruefully, the authors tell us that even though some "efforts have been made to build healthy buildings," no "coherent view [...] of how a healthy building should be constructed, and of how such a building should be used and maintained in order to remain healthy," has as yet been conceived.[15]

Nevertheless, they have many things to say about indoor climate and light, about air recirculation and humidification, about building sites and materials, as well as about design, operation and maintenance, considerations that will, hopefully, contribute to minimizing the health hazards for those who stay or live in those buildings. "[T]rue healthy building" emerges as a prevision of a distant, much desired future, and as a down-to-earth attempt at providing current technical guidelines, sometimes even specifications, for the construction of buildings that do not cause illness or disease. On both counts, the metonymical phrase "the healthy building" is a celebration: of current technical skills and technological know-how, and of a prospective future, which is, so far, little more than a grand vision.

But healthy houses and healthy buildings may form part of a larger, more ambitious project, namely, healthy cities. Such a project was in fact run by the World Health Organization in the late 1980s.

The Healthy Cities Project of the WHO was conceived broadly, beyond the concerns of traditional health care. One leading official suggested that the inhabitants of every larger city, jointly and severally, "develop a vision of what the preferable future should be like, then develop that into an easily understood and plausible scenario and use that as a goal." More specifically, the future health of their cities is at stake, and the task is to envision cities that would be "as healthy as possible," or "ideally healthy."[16]

Vision and scenario are supposed to be followed by "environmental scanning" and a "strategic plan." And the strategic plan, in its turn, should feature "a set of strategic priorities," such as the reduction of "inequities in health opportunities," the creation of physical and social environments that support health, the strengthening of the community's ability to take health-protecting action, the development of individual health-promoting skills, and the reorientation of health services "towards health promotion, disease prevention and community-based care."[17]

No city is itself healthy. In other words, no city is literally healthy. A city is healthy only insofar as it is built, organized and run in such a way that the health of its citizens is supported, protected, and improved. The metonymic relation between a healthy city and the health of its citizens is therefore causal. But it is also a whole-parts relation; we care about the health of people who live in cities.

The WHO Healthy Cities Project presented "healthy cities" as a through-and-through metonymical phrase: healthy cities are cities in which the physical and social environment, the political and economic leadership, and the health care sector, more narrowly conceived, conspire to promote and to optimize the health of all those who live there. "Healthy cities" is a catch phrase that summarizes a lot: from lofty visions, which are themselves supposed to promote health, to specific aims and actions which are directly causally related to improved health among those who live in cities.

A project like that of the WHO need not be too particular about what "health" means. The project may, confidently, draw on the vague and multifaceted notion of health with which we are all familiar: health as the absence of disease; as a sense of strength and well-being; as fitness for work and leisure; as ability to realize cherished goals in life; and so on. We all know, vaguely, what health is all about, even though we may, with good reason, be unable to define it.

6. Metonymy and/or Metaphor: From a "Healthy Climate" to "Healthy Respect"

We have noted that "healthy" is often used metonymically, as in "healthy cities" and "healthy houses." The atmosphere and the sanitary conditions of cities and houses are, in obvious and less obvious ways, causally related to the health of those who live in those cities and houses. And when phrases such as "healthy cities" and "healthy houses" are used in lectures and reports, they celebrate a real or imagined knowledge of what these causal relations look like, as well as a technological know-how related to city planning, architecture, and construction.

If we look at the phrase "a healthy climate," a similar metonymic relation between cause and effect is suggested—only you cannot plan, construct, and accomplish a healthy climate.

A "healthy appetite" might be considered a similar case of metonymy. In some people, and on some occasions, appetite and eating indicate health even as they further it. "Appetite" and "health" then inhabit the same semantic domain, and a metonymic phrase may be both efficient and economical in suggesting how they are related.

On two other interpretations of the phrase "a healthy appetite," "healthy" is a metaphor: 1) a healthy appetite may be a normal, average and inconspicuous appetite, an appetite that does not stick out—just as being healthy may be felt to be the natural, inconspicuous and uneventful way of being; and 2) a healthy appetite may be an appetite that is noticed by others, a remarkable, conspicuous, if not voracious, appetite—just as a healthy human body may be conspicuous by its vigorous appearance. Both metaphors draw on similarities between the appearance of health (in the literal sense) and the appearance of people who betray their appetite by their eating habits. It is notable that these two opposite metaphorical senses of "healthy" testify to the notorious vagueness of the literal sense of "health" and "healthy."

If we turn to phrases like "a healthy curiosity" and "a healthy respect," we will find that metaphorical interpretations of "healthy" tend to override metonymical ones.

A healthy curiosity is a curiosity that is alert as well as desirable, desirable because it is deemed desirable to be curious, though presumably not about other people's private lives. Curiosity may be considered healthy if it strikes a balance between too much and too little. These metaphorical interpretations of "healthy" are more likely to come to mind than their metonymical rivals, which would convey the idea that curiosity leads to health, or that being curious is part of what is meant by being healthy.

The phrase "healthy respect" is even more clearly metaphorical rather than metonymical. A healthy respect may be a normal, natural and desirable respect for other people, their feelings, preferences, and opinions. It may even be a respect above the normal or average, and so a particularly strong and virtuous form of respect. A metonymical interpretation of a "healthy respect," on the other hand, would convey the idea that respect leads to health or is part of what is meant by health. That idea is a bit far-fetched, though not outright silly.

A similar analysis appears appropriate if we turn to a "healthy attitude." A healthy attitude evokes approval, because it is deemed desirable, or even a virtue. That is the metaphorical interpretation. As a metonym, a "healthy attitude" would indicate that the attitude in question furthers health. And that is perhaps a bit far-fetched.

7. Metaphor and Metonymy in a "Broken Health"

In a report on the life history of a man of the 20th century, we might find the following statement: "After ten years of hard labor, his health was broken."

Searching this statement for figures of speech, "hard labor" is our first catch. That is a set phrase, in which "hard" is, no doubt, a metaphor. The whole phrase, "hard labor," may, in some contexts, be a metaphor, though hardly in this one.

Leaving the set phrase "hard labor" aside, we notice that the statement as a whole is, obviously, metaphorical, since no one's health can, literally, be broken. Still, we may wonder: Which of the words employed turns the whole into a metaphorical phrase? The two candidates are: "health" and "broken."

Let us begin with "broken" as metaphor. Presumably, most people asked to single out a metaphor from a "broken health" would choose "broken." "Health" is then taken to be as literal and, in spite of its vagueness, straightforward as it is supposed to be in common usage. The term may refer to a feeling of bodily strength, or to the absence of disease, but either way it is not a metaphor. "Broken," on the other hand, is borrowed from a down-to-earth physical context: cups and plates break, as do trees and legs, when exposed to physical force of a certain kind and magnitude. But again, no one's health literally breaks.

Only when borrowed from the everyday world of physical force into a practical and semantic domain that makes frequent use of abstractions—such as that of medicine and health care—does "break"/"broken" turn into the telling metaphor of a "broken health." We may imagine a heavily overworked man, exhausted, weak, and bent, his health gone. That man, as represented in the imagination, is similar, in some respects, to broken physical things like trees, cups, and tools. The integrity of either is lost. "Broken" as metaphor draws on that similarity. Its message is clear: No health is left, none whatsoever; health is irretrievably lost and permanently absent.

Let us now consider "health" as metaphor, and "broken" as a word used literally. "Health" must then stand for something that is literally broken. What might that be? "Hard labor" provides a clue: it is the body that, sooner or later, breaks down under the pressures and privations of hard labor. Or, even more literally, it may be the legs and bones of the body that break. After ten years of hard labor, a human body may be physically, and literally, broken. Perhaps that is what happened in this case, only that the reporter did not say it in so many words. But what he did say is no less evocative: After years of hard labor, his health was broken.

But if "health" stands for the body, a human body, is it then a metaphor? Abstractions—and "health" is an abstraction from everyday experience and from medical practice—rarely make good metaphors. Besides, "health" and "body" do not belong in separate linguistic-semantic domains. Health is typically sought and achieved by way of the body or organism, which is taken care of, supported, and mended, as the case may be. In other words, "health" and "body" inhabit the same semantic domain, a fact that militates against the generation of that tension between semantic domains which is of the essence of metaphor. We are led to the conclusion that though "health" may stand for the human body, it cannot stand for it metaphorically. In other words, "health" is hardly a metaphor at all.

Still, "health" is here a figure of speech, albeit not a metaphor. We know that, since health does not literally break. The relation between health and the body is not metaphorical but metonymical: "health" stands for the body as one of its aspects, or as a net result of its constitution and inner events and processes. Whether regarded as an aspect or as an effect of the human body and organism, "health" is a metonym. And it works as metonyms often do, namely, as short for something more complex. In this case, "health" is short for everything in the human body that can, and under unfavorable circumstances does, break.

Finally, to see how "health" in the literal sense and "health" as a metonym are related, I will again focus on that poor man whose health has been broken after ten years of hard labor. He may have contracted several diseases, and he is no doubt weak, with little or nothing left of the sense of strength that he once, we may surmise, had. His legs and bones may be fractured, or badly healed. To say that he is in poor health is an understatement. Rather, his health is gone, totally and irretrievably lost. And that is precisely what is being reported: his health is broken.

"Health" may then, to begin with, be taken in its ordinary, general, vague and unstable sense: as bodily strength; as vitality; as fitness; as absence of disease, and so on. When all this is gone, health is gone, health is broken. "Broken," in its turn, is a metaphor that draws on an imagined similarity between the sad spectacle of a man with a broken health and broken—literally broken—physical things like cups, trees, and tools. But then, again, we can see that "health" may be considered a metonym, not to compete with, but to complement its ordinary, vague, and unstable sense. Because "health" is then short for all those parts and mechanisms of a man's body which normally make him strong, but which can also literally break and make him look like broken physical things such as tools, trees, and toys. "Health" in the literal sense invites, when coupled with the metaphor "broken," "health" as a metonym.

8. A Proverb: "A Nation's Health is a Nation's Wealth"[18]

The proverb "A nation's health is a nation's wealth" exhibits "health" as a metonym, but possibly also as metaphor.

One interpretation of this proverb might run: If the citizens of a country are healthy, they will work hard and diligently, which will lead to general prosperity in that country. "A nation's health" then stands, metonymically, for the health of its individual citizens, most citizens, who will be able to accomplish great things together, accumulating wealth. Here the whole stands for its constituent parts, which is one of the linguistic mechanisms by which a metonym is formed. We should also notice that since "health" retains its ordinary, general, and vague meaning, it is, strictly speaking, the phrase "a nation's health," not "health" taken by itself, that stands metonymically for the health of its citizens.

Taking that metonymical explication of "a nation's health" as our point of departure, we may proceed with an interpretation of the proverb that recognizes that "is" appears to be a word of logical identity. We will then be able to notice an intriguing mixture of, and possibly rivalry between, metonymy and metaphor.

To begin with, we may notice a cause-effect relation between healthy citizens and the wealth or prosperity they manage to attain. A nation's health, explicated as the health of its citizens, stands for the wealth and general prosperity that that health will enable them to achieve. Cause stands for effect. In other words, the relation between health and wealth is metonymical. Notice that this metonymy is founded on the metonymical reading of "a nation's health" previously noted: a nation's health is the health of its individual citizens, jointly and severally. Through hard and diligent work, their health, in the literal sense of physical and mental strength, will achieve nationwide success in terms of wealth and general prosperity.

However, insofar as "health" and "wealth" belong in two distinct semantic and pragmatic domains, a note of metaphor would be unavoidable when it is said that a nation's health is its wealth. If we stick to the overall interpretation here under consideration, which reads "wealth" literally, "health" as metaphor turns out to stand for strength, greatness, and, generally, for that which is, and ought to be, valued highly. What the proverb says, then, is that the *wealth* of a nation ought to be so valued; a nation's wealth stands for its true greatness.

The metaphorical interpretation of "health" appears to be more farfetched than the metonymical one. Why? Mainly, I gather, because of the exact wording of the proverb: "health" comes first, with "wealth" only in the second place. This arrangement puts emphasis on health, which appears to be more interesting in its own right than wealth, whereas wealth appears to be no more than an explication of health. The presumption is that a word that is supposed to be defined or explicated is used in its literal, not some metaphorical, sense. If the wording and order of the proverb were reversed, these relations between health and wealth would also be reversed. In the proverb: "a nation's wealth is a nation's health," the presumption would be that "wealth" is to be taken literally, whereas "health," which explicates "wealth," might be a metaphor, signifying strength and greatness.

Now, we might ask: How come that "health" in this proverb (in its original wording) can be interpreted either as a metonym or as a metaphor (though preferably as the former)? This has to do with the generality and vagueness of "health" taken literally. "Health" lends itself to several interpretations and meanings. You may use "health" without bothering much about specifications and precision. Or you may choose to be as specific and precise as you possibly can.

Sometimes the addressee-interpreter feels invited to supply further determinations, which will render "health" less general and vague. Sometimes he or she will just have to wait and see what "health" means; future events and human interaction will supply determinations. If the interpreter is inclined, or chooses, to emphasize the vagueness and instability of "health," he or she will be relatively more inclined to take it for a metaphor, signifying strength and, generally, that which is, and ought to be, valued highly. But again, the wording of this proverb as we have it suggests a metonymical interpretation, where "health" is closely related to human bodies and organisms and their abilities to accomplish things.

Yet another interpretation of the proverb might run: The health of the citizens of a country is itself a species of wealth; moreover, health being the most valuable thing for a human being, it takes precedence over wealth in the more ordinary sense of the word (abundance of money, property, or other valuable things). Here "a nation's health" is, again, a metonymical phrase, which stands for the health of individual human beings, whereas "wealth" assumes a metaphorical character, standing for that which is, and ought to be, valued above other things. The message of the proverb on this interpretation, then, is: In any nation, the health of its citizens ought to be regarded as the most important form of wealth, more important than money, property, possessions.

Finally, we may consider an interpretation of the proverb that turns "health" into a political metaphor that stands for social order, peace and harmony domestically, and for power, influence, and a good reputation in the international arena. Such an interpretation might run: If you wish to measure a nation's true greatness and dignity, look at its level of social order and harmony, as well as on its international reputation, power, and influence. "Health" as metaphor here draws on the similarity between a body or organism that is strong and harmonious in its inner workings and outward appearance, and a state, nation, or body politic characterized by social order and harmony domestically, and power and influence internationally.

On this interpretation, "wealth" too is turned into a metaphor, standing for true moral greatness and dignity. These are values that evoke respect and admiration, and sometimes envy. So does wealth in the ordinary sense of the word. "Wealth" as metaphor draws on that similarity.

9. "The Health of the Scientific Community"—Possibly Metonymy, but Preferably Metaphor

The American philosopher Charles Sanders Peirce once wrote: "the health of the scientific community requires the most absolute mental freedom."[19] This is a beautiful statement, amounting to a moral rule and maxim.

What could be meant by the phrase "the health of the scientific community" here? And what did Peirce mean to say by using it?

On a metonymical reading, a concern for "the health of the scientific community" is a concern for the individual members of that community, their health and well-being. The predicate of Peirce's statement, "requires the most absolute mental freedom," is consistent with such a reading and renders it plausible. Scientists must be allowed to feel, react, and think freely, that is, under no constraint (except their own), in order to remain healthy, not least mentally; because their frustration if denied such freedom will cause illness, if not outright madness.

That metonymical reading makes sense. Yet, it appears a bit awkward and far-fetched. Moreover, the medical (preventive or therapeutic) ring of Peirce's statement as metonymically interpreted is distinctly un-Peircean. Peirce cared for the health of scientific inquiry and reasoning, not for that of scientists. This brings us to a metaphorical reading of his statement, "the health of the scientific community requires the most absolute mental freedom."

On a metaphorical reading, the health of a scientific community is a dynamic display of what constitutes that community as scientific: the pursuit of truth through sustained inquiry into the nature and workings of things. In other words, a healthy scientific community is simply what it is: scientific, a scientific community. Its integrity depends of this being so. Moreover, the indicative mood of the statement issues an imperative: Be what you are. The positive value judgment inherent in the word "health" is borrowed into the sphere of scientific inquiry. A healthy scientific community, which is a truly scientific one, is, no doubt, a good and praiseworthy thing. It is worth pursuing, and, once established, it is worth defending.

"Health" works as metaphor in this context partly because of its positive value-ladenness, but also because health in the literal sense is associated with an idea of integrity. The integrity of the scientific community, which is intellectual, may be represented to the mind's eye parallel to the integrity of a healthy human body, which is physiological. The similarity which then emerges between two forms of wholeness or integrity turns "health" into a useful and enlightening metaphor: "the health of the scientific community."

When Peirce goes on to say that "the most absolute mental freedom" is required for the health of the scientific community, this can be read as a further explication of what he takes genuine scientific inquiry to be.

Two aspects of mental freedom can be distinguished, one negative and one positive. Negative freedom is freedom from external constraints and authoritarian dictates that interfere with the process of inquiry, by stipulating the ideas, terminology, hypotheses and methods that, supposedly, ought to be used by the scientific inquirer, or by banning the ideas, methods, etc. that a given inquirer wishes to employ. Positive freedom for a scientific inquirer is a freedom to make the most extensive use of whatever mental powers he or she happens to be endowed with, of feeling, perception, reaction, memory, imagination, reasoning, or judgment, in order that no seminal hypothesis, no pregnant observation, and no hidden implication, be lost.

"[T]he most absolute mental freedom," which is Peirce's concern and forte, would include both aspects of freedom, and eminently so. The "health of the scientific community," which is the integrity and dynamic display of genuine scientific inquiry, requires such extensive and substantive freedom. The health metaphor in Peirce's maxim is made the focus of an entire philosophy of science, a cognitive ideal, and a methodological program.

10. The Moral Message of Health Metonyms and Metaphors, Respectively

Metaphors commonly appear to be more interesting than metonyms. Metaphors are small pieces of food for thought, which nourish the mind even as they cause pleasure; whereas metonyms are vehicles of word economy, which allow the mind to take shortcuts to destinations that are well-known, familiar, even commonplace. Compare the metaphorical phrase "the health of the scientific community" with the metonymical "health food," and notice the contrast between an appetizing metaphor, fraught with unspecified meaning, a surplus of meaning, and a pedestrian metonym, which evokes the consistent picture of tins and cans and packs of not very appetizing, though supposedly healthy, food.

Anything that we believe will further health is ripe for metonymy: a healthy diet; healthy exercise; a healthy rest; a healthy way of life; healthy working conditions; and so on. Such metonyms may be extreme abridgments of advanced scientific hypotheses and theories of health promotion. Or they may derive from mere guesswork, hearsay, or current mores. Either way, the perspective on health and health promotion is straightforward.

Health metaphors convey their message in a more roundabout way. You do not just grasp their message; instead, you have to catch on to it, and to do that you may have to think, ponder, ruminate. Consider, for example, some of the metaphors that attracted our attention earlier in this essay: "healthy respect"; "healthy curiosity"; and "the health of the scientific community."

However, their stylistic and semantic differences notwithstanding, health metaphors and health metonyms share one characteristic: Both commonly convey an implicit moral message, namely, that health is a good thing that is worth pursuing. Health is good when you possess and enjoy it. But health is also a good and a goal to be pursued and attained in yourself and others. Hence, health is both laudable presence and demanding project. And that project is, to some extent, a moral one: health as a goal of virtuous action.

Most people take for granted that they have some obligation to take care of and to promote their own health, and possibly also that of a few others, such as members of their family. This paves the way for all those metonyms that convey ideas about health-advancing living conditions, actions, and ways of life: a healthy climate; a healthy diet; healthy exercise; and so on. These and similar metonyms suggest moral messages premised on the following obligation: You ought to take care of, and to enhance, your health, and possibly that of others too.

The messages of health metonyms are usually straightforward do's, premised not only on a basic moral obligation, but also on simple causal relations: Some foodstuffs promote health, whereas others promote its opposite; some sporting activities advance health, whereas others cause injuries and hence ill health; and so on. Scientific investigations may prove that some suggested causal relations are not so simple and straightforward. But that does not necessarily influence usage. A popular health metonym may live on in spite of its having been proven spurious. It has then acquired a life of its own. The sober, critical scientific spirit may prove impotent against such linguistic autonomy.

On comparison, the moral messages of health metaphors tend to be more sophisticated, just as metaphor is a more sophisticated figure of speech than metonymy. Health metaphors commend health, but metaphorical health is regularly more elusive and thought-provoking than metonymical health. You have to figure out what sort of "health" a health metaphor asks you to promote. Consider, again, "a healthy respect;" "a healthy curiosity;" and "the health of the scientific community."

NOTES

1. Louis Goossens, "Metaphtonymy: The Interaction of Metaphor and Metonymy in Figurative Expressions for Linguistic Action," in *By Word of Mouth: Metaphor, Metonymy and Linguistic Action in a Cognitive Perspective*, ed. L. Goossens, P. Pauwels, B. Rudzka-Ostyn, A-M. Simon-Vandenbergen and J. Vanparys (Amsterdam: John Benjamins Publishing Company, 1995), pp. 159–174.
2. *Ibid.*
3. *The Concise Oxford Dictionary of Current English*, Eighth edition (Oxford: Clarendon Press, 1990).
4. *The American Heritage Dictionary of the English Language*, Third edition (Boston: Houghton Mifflin Company, 1992).
5. *Ibid.*
6. *Ibid.*
7. *The Concise Oxford Dictionary of Current English.*
8. *The American Heritage Dictionary of the English Language.*
9. *Ibid.*
10. *Longman Modern English Dictionary* (Harlow: Longman, 1976).
11. *The Concise Oxford Dictionary of Current English.*
12. *Longman Modern English Dictionary.*
13. *The American Heritage Dictionary of the English Language.*
14. Nina Dawidowicz, Thomas Lindvall, and Jan Sundell, eds., *The Healthy Building* (Stockholm: The Swedish Council for Building Research, 1988).
15. *Ibid.* p. 3.
16. Trevor Hancock, "The Healthy City: Utopias and realities," in *Healthy Cities*, ed. John Ashton (Buckingham: Open University Press, 1992), pp. 22–29, on p. 23.
17. *Ibid.* pp. 26–28.
18. Wolfgang Mieder, Stewart A. Kingsbury, and Kelsie B Harder, eds., *A Dictionary of American Proverbs* (New York: Oxford University Press, 1992).
19. C. S. Peirce, *Collected Papers*, ed. Charles Hartshorne, Paul Weiss, and Arthur Burks (Cambridge, Mass.: Harvard University Press, 1931–58) 2.220.

Three

ON THE EVOLUTIONARY CONCEPT OF HEALTH: HEALTH AS NATURAL FUNCTION

Lennart Nordenfelt

1. Introduction

A controversy exists concerning the nature of health and illness in many health-related disciplines. Two mainstreams have appeared in the arena. Proponents of the first stream of thought, which is medical or biological in nature, typically claim that the concepts of health and illness can be treated as purely biological concepts. "Health" and "illness" are taken to be biological concepts in the same sense as "heart" and "blood pressure" are biological concepts. Nothing is subjective or evaluative about these concepts, it is claimed. I will call this view the naturalistic line of thought.

The other mainstream takes a completely opposite position with regard to these matters. According to its proponents, the normativists, health and illness are intrinsically value-laden concepts. These concepts cannot be totally defined in biological or psychological terms, if these terms are supposed to be value-neutral. To say that somebody is healthy partly means that this person is in a good state of body and mind.

My position is on the normative side (1995, 2000, 2001). I have argued extensively against naturalism, in particular against its most celebrated version, presented by the American philosopher Christopher Boorse. That theory is based on the general idea that the healthy is the natural in the sense of being biologically normal. According to Boorse's explication, biological (or medical) functional normality is relative to the goals that "physiologists seem to assume, namely, individual survival and reproduction" (1997, p. 9).

But a rival naturalistic line of thought entered the arena during the 1990s. This view places the notions of health and illness within the context of evolutionary theory. The notion of biological dysfunction is crucial here. According to this theory, a condition must be a dysfunction or be caused by a dysfunction in order to qualify as an illness. The major proponent of this evolutionary line of thought, Jerome C. Wakefield, has promoted his ideas mainly within the current debate on the nature of mental disorders, and his theory has therefore not become so well-known in the field of the general philosophy of medicine. His theory, if valid, could, however, also be applicable for the characterization of somatic health and illness.

Wakefield's basic supposition is similar to Boorse's; namely, the ancient concept that the healthy life is in some sense the natural life, life that is in accordance with the biological nature of humankind. On a superficial level, this idea appears quite reasonable. A vague metaphysics of this kind is probably embedded in the Western layperson's understanding of life. When we first try to theorize about health, it is "natural" for us to commence with this intuitive idea. I myself started off as a naturalist, but later became convinced that this philosophy must be a dead end.

That the metaphysics of the ordinary human being is vague can be illustrated by the two directions in which an understanding of the natural can be interpreted and refined. Boorse and Wakefield develop their ideas differently. Boorse abstains from referring to biological evolution (although some commentators put him in this camp; for instance, Wakefield (1992a), and George Khushf (2001)) in making his characterization of health. Boorse (1976, p. 75) explains his relation to evolutionary accounts in the following way:

> It is operational explanation, not the evolutionary sort, with which physiology has traditionally been concerned. Confronted with inordinate complex organisms that achieve some goals in a remarkable range of conditions, physiologists set out to study them as you would study an alien artifact; that is by determining what contribution, if any, each part makes to the operation as a whole.

Boorse's theoretical characterization of the natural is exclusively related to the present context: The healthy organ is one that makes a statistically typical contribution to the two main biological goals, namely, survival of the individual and survival of the species. Wakefield, however, underlines the historical evolutionary context in explicating the notion of a natural function, which is the crucial notion for his characterization of health. I will devote the bulk of this paper to penetrating this notion and its viability for a reasonable analysis of medical notions such as health and illness.

The starting point for Wakefield's analysis is the state of psychiatric theory around 1990. At that time, much controversy existed concerning the status of the *Diagnostic Statistical Manual of Mental Disorders (DSM)* and the criteria that the constructors of this manual used in selecting conditions suitable for registration in it. Since its third edition, *DSM III* (1980), this manual has included a general definition of mental disorder:

> Each of the mental disorders is conceptualized as a clinically significant behavioral or psychological syndrome or pattern that occurs in an individual and that is associated with present distress (for example, a painful symptom) or disability (impairment in one or more important areas of functioning) or with a significantly increased risk of suffering death, pain, disability or an important loss of freedom (*DSM IV*, 1994, p. xxi).

Many theorists have expressed the complaint, however, that this definition has not sufficiently distinguished pathological from non-pathological disorders. This, some critics claimed (Kutchins and Kirk, 1997), was partly because the definition itself needs interpretation and amendment. Wakefield has adopted this task of interpretation in several seminal articles during the 1990s.

Wakefield (1992b) starts with an improved and simplified version of the DSM definition: A mental disorder is a mental condition that (1) causes significant distress or disability, (2) is not merely an expectable response to a particular event, and (3) is a manifestation of a mental dysfunction. Wakefield thinks that (1) and (3) are the crucial and valuable elements in a definition of mental disorder; (2) is, he thinks, redundant or misleading. He thinks (I believe correctly) that some instances of mental disorder are expectable. If a person is exposed to a sufficient amount of stress and damage, then the expected outcome is disorder, whether physical or mental. The expected outcome cannot be a disqualifier. Harm and dysfunction, on the other hand, are essential elements in Wakefield's characterization of mental disorder.

This means that Wakefield's position in the theory of health is complex. His position has naturalistic and normative elements. A mental disorder is a defect from the point of view of natural order; it is a harmful state of affairs. In short, a mental disorder is a harmful mental dysfunction. More expanded: "A disorder exists when the failure of a person's internal mechanisms to perform their functions as designed by nature impinges harmfully on the person's wellbeing as defined by social values and meanings" (Wakefield, 1992a, p. 373). But the dysfunction requirement "is necessary to distinguish disorders from many other types of negative conditions that are part of normal functioning, such as ignorance, grief and normal reactions to stressful environments" (Wakefield, 1992b, p. 233).

Wakefield says rather little about the harm aspect of the mental disorder. We should translate this aspect, he says, into a list of specific and recognizable kinds of harm, especially stress and disability. Such a proposal he shares with almost all normativists within health theory. However, in one respect he takes a stand in favor of a social interpretation: "only dysfunctions that are socially disvalued are disorders" (Wakefield, 1992a, p. 384).

Wakefield's analysis focuses on his attempt to define the notion of natural function and apply it to the area of mental disorder. This is also where I will focus my analysis. The main questions are: What is the meaning of "natural function" and its negative derivative "dysfunction," and what role can these notions play in the theory of health? I will generalize the discussion to cover health and illness in general, not just mental disorder.

2. The Notions of Dysfunction and Natural Function

Wakefield's preliminary specification of the notion of dysfunction is the following:

> A dysfunction exists when a person's internal mechanisms are not able to function in the range of environments for which they were designed. Thus, one can construct a test for dysfunction by specifying an environment in which the function is designed to manifest itself; if the function fails to be manifested in that environment, there is likely a dysfunction (Wakefield, 1992b, p. 243).

A dysfunction, Wakefield says, should be related to the natural function of an organ in an organism. Every organ has, we intuitively think, a natural function. The most celebrated example is the function of the heart, which is taken to be the circulation of blood in the body of the animal in question. Sometimes an organ has more than one natural function; the liver, for instance, will both activate the body's glycogen synthesis and purify the blood from poisons. But how should we analyze this notion of natural function? Literature on the subject includes attempts that are only slightly different from one another. I will focus only on Wakefield's version.

A careful analysis of the concept of natural function, says Wakefield, leads to the conclusion that the most viable approach to it is based on the notion of evolutionary design.

> Certain mechanisms are naturally selected because of the beneficial effects that the mechanisms have on the organism's fitness, and those beneficial effects are the natural functions of those mechanisms. The natural function, then, is not just any benefit or effect provided by a mechanism but a benefit or effect that explains through evolutionary theory why the mechanism exists or why it has the form that it does (Wakefield, 1992b, p. 236).

Wakefield then asserts that this concept of natural function is a purely scientific one, although he concedes that the process of discovering what is natural or dysfunctional may be extraordinarily difficult. On this point Wakefield allies himself with Boorse, who also underlines the purely scientific or value-free nature of his concept of health. The question of whether a value-free scientific notion of natural function exists is, at present, the focus of a lively discussion. (See *The Journal of Abnormal Psychology*, 1999, No. 3, and

Philosophy, Psychiatry, and Psychology, 2000, No. 1. I will bypass this specific question here, although I will have to touch it indirectly.)

My main questions are: Can we find a coherent and reasonable concept of natural function? And if we can, does this concept have anything to do with a viable notion of illness or disorder? First, a few ontological points. The notions of function, functional ability, and the exercise of a functional ability are frequently confused. Often, these terms are used interchangeably.

An organ may have a function; to say this is to make a statement on the species level. The function of the human heart is to reach some state of affairs or to keep a process going. The function then relates to a result of the working of an organ. The heart of an individual human being may have an ability to realize this function, or it may not. If not, then it is dysfunctioning. An individual organ may or may not at a particular point exercise its functional ability. Some organs exercise their functional ability all the time, as the heart does. Other organs, such as the stomach or the kidney, exercise it only at some times.

I will now go a little into the basic idea of a natural function, beginning with the heart. The heart pumps blood. The result of the pumping, the distribution of the blood, is the natural function of the heart in a human being. What does this mean? The heart has this function, says Wakefield, because the human heart, like the hearts of all animals, was designed in this way and for this purpose.

But design is metaphorical talk, admits Wakefield. The terms "species design" and "selection" are metaphors, as are several other terms in this area of discourse. They are chosen in analogy with the human case of artifacts, where a human being intentionally designs and selects. No such thing, or any analogous assumption related to the intentions of God, is assumed in the theory of evolution. Wakefield and others who put forward the evolutionary argument underline this point.

The idea of design must have a non-intentional interpretation in the field of biology: a causal interpretation. A particular causal history that crucially involves the distribution of the blood lies behind the existence of the heart in a human being. At some stage in the early history of humankind, individuals quite similar to us existed. They had hearts and lungs that worked in a way that was similar to the way hearts and lungs work in us. These hearts and lungs contributed to the survival of those individuals. Since human beings had several properties that, together with the work of hearts and lungs, made them fit to survive as a species (given the environments existing up to our time), we still have human beings on the earth. All human individuals today have a heart that pumps blood and thereby distributes it around the body. Since the similar work of hearts long ago contributed to the survival of human beings, we can designate the distribution of the blood to be a natural function of the human heart.

Here is a straightforward causal story. We find no mention of human values, nor of intentionality by God or human beings, just strict causality, strict natural science. This description follows a standard analysis of functions. A *locus classicus* in this respect is Larry Wright's *Teleological Explanations* (1976), p. 81:

The function of X is Z iff:

(i) Z is a consequence (result) of X's being there,
(ii) X is there because it does (result in) Z.

To illustrate: The function of the heart is to distribute the blood in the organism. The distribution of the blood is a consequence of the heart's being in place in the body. The heart exists because the distribution of blood has been a necessary causal condition of the survival of all the ancestors of the present members of species X, as well as of these individuals themselves. (For a penetrating analysis of this traditional notion of function, see Boorse, 1976.)

So much for the introduction of the idea of a natural function; the specification of a dysfunction is then easy. An organ or a mental faculty has a dysfunction, or is dysfunctioning, if and only if this organ or faculty does not (completely) fulfill its natural function as specified above. The dysfunction is a property of an individual within a species. It is a *contradictio in adiecto* to talk about a species that is dysfunctioning. This observation has some bearing on the discussion about the relation between dysfunction and disorder.

I will now draw some further crucial conclusions from the notions of natural function and dysfunction.

3. On the Identification of Natural Functions and Dysfunctions

A. Difficulties with Regard to Identifying Natural Functions

Not all so-called functions are as easily delineated as the distribution of blood in the case of the heart. This function is salient in the sense that the distribution of the blood is necessary for the survival of the organism at all moments of the life of the organism. If the heart stops pumping, the organism soon dies. The distribution of the blood is a strictly necessary condition of the individual's survival today. We can then plausibly presume that this was a necessary condition for the survival of human beings and previous anthropoids long ago. And if something is a necessary causal factor in respect of the existence of every individual belonging to a particular species, it is a partial causal factor in respect of this species's continued existence today.

But what about other functions? I will first test a different case where we are perhaps also inclined to talk about a natural function. We may, for instance, argue that the function of the human hand is to grasp things. But this case is not as easy as the case of the heart. An individual will not immediately die if such an ability diminishes or ceases to exist. We cannot therefore say with the same certainty as in the case above that the function of grasping in the past causally contributed to the existence of human beings today. We may speculate with some plausibility that the grasping of things has been crucial for the survival of the human species. But what we are doing here is only offering a plausible speculation. (Even more difficult is the test with regard to some mental faculties such as cognitions and emotions. I will briefly return to these faculties later.)

But what should we say about such parts of the body where we do not have any initial intuitions about a natural function? This actualizes a central problem concerning the idea of historical evolution. This idea focuses on what functions were crucial to human beings in the past—what made us survive in the past, given the standard circumstances in the past, or even given some special circumstances in the past.

Consider the following: It may have been the case that the human species was dying and almost extinct ten million years ago. Human beings (or our anthropoid ancestors) were, according to this assumption, not at all fit for the climate and for the general ecological situation at that time. The so-called natural functions of those human beings were not viable in that context. But because of a climate change, a pocket population was rescued and these human beings could start building up a viable population again.

Plausible speculations concerning the benefits of some functions and faculties for the survival of human beings today cannot be immediately applied to establish something's status as a natural function. We can see that today it is necessary for a human being to grasp things, or to reason about things and to memorize things, to take two mental examples. The ability of memorizing things is of particular importance today when we, for cultural reasons, are forced to remember a large quantity of number combinations. But this ability need not have been so crucial in ancient times.

But the fitness of a function today is not the criterion of its being a natural function according to Wakefield's definition. (He is quite explicit on this point in his criticism of some imperfect evolutionary accounts of health and illness; 1992a, p. 379.) We may get a clue for identifying a natural function by considering its survival value today. The criterion is its fitness through history; the function in question may be of limited or no use today and still be a natural function. Probably some physical capacities exist that are more or less superfluous today, but were quite relevant long ago in the old human societies of hunters and fishermen. This, then, marks the main difference between Wakefield's notion of natural function and Christopher Boorse's notion of normal function. Boorse's is completely focused on the beneficial strength of the function today. He says that the normal (read: healthy) function is the one that makes a typical contribution to the individual's survival, or the survival of the species, today.

Delineating the natural functions of the human organism (as well as of all other organisms) is not only a difficult affair; it is, strictly speaking, impossible. With the exception of the most salient functions, such as the main functions of the heart and the lungs, we are, and must remain, too ignorant to tell whether a particular present disposition of an organ or a faculty should be labeled a natural function. This is not a promising state of affairs as concerns the usefulness of the notion of natural function as a criterion of health.

B. On Determining the Proper Degree of a Natural Function

Similarly, we might wonder how the historical evolutionary perspective can help us determine what is the natural or proper degree of an alleged human function. How much blood should the heart pump? How well should the hand grasp? How much should someone be able to memorize? How much sorrow should we feel when a relative dies? What resources, other than statistical ones, do we have to determine what is a proper degree?

Since the environment, in terms of climate and other natural conditions, has changed so considerably over time, the demands on people have changed greatly. At one time, human beings in the north had to withstand cold and be able to survive on few nutrients. At some other time, human beings may have been forced to compete with other animals in predation. At some further stage, human beings may have lived quite sheltered and with an abundance of food. The requirements have varied over time and place. Assume that at some time, t, a function, F, to degree 10 was necessary for the survival of a population, while at time t1, F was necessary to degree 8, and now that function may be necessary to degree 4 for the survival of the present population. What is then the "proper" natural function as quantitatively specified? Which historical stage should we choose for this decision? We can hardly use the present level. That goes against the importance of the evolutionary perspective: The function should be something that has been historically selective. But what segment of history should we choose? The choice appears to be arbitrary.

I will on this point quote one of the most prominent evolutionary biologists, Ernst Mayr (1982, pp. 483–484):

> Organisms are doomed to extinction *unless they change continuously* in order to keep step with the constantly changing physical and biotic environment. Such changes are ubiquitous, since climates change, competitors invade the area, predators become extinct, food sources fluctuate; hardly any component of the environment remains constant.

I wish to discuss a plausible argument that we can give to this observation. According to this argument, the "real" functions do not change—or the proper degree of the real functions does not change—over time. The changes occur instead on a more superficial level. Take the human heart as an example. The heart pumps blood. It is elementary that the pulse varies a lot with the environment and with the activity of the subject. We cannot therefore say that a particular frequency of the beating of the heart is the "natural" frequency. The frequency in a way compensates for changes in the environment in order to attain the crucial result of distributing blood, and thereby oxygen, to all other vital organs and tissues of the body in the required manner. The pulse is not in itself a "real" function according to this reasoning. The real function is the distribution of oxygen to other organs and tissues. This may be performed equally well in different kinds of situations.

The problem with this argument is that we can observe equally as many possible differences on the higher level, namely, regarding the distribution of oxygen. The environment does not just require different pulses, but also different levels of oxygen for different purposes and different contexts. The blood-distribution also becomes disqualified as a "real" function. We must look for the real function on an even higher level of the organization of the human being. But also on that level we can find differences and no "real" functions.

The search for the real function will continue until we end up on the completely abstract level, where we say that the hearts in the different environments make equivalent causal contributions to the survival of the individuals in question. This may be true, but we have then completely capitulated in our endeavor to find functions on the concrete level of the organs where we wanted to find them for the purpose of detecting particular dysfunctions.

4. On the Scope of the Notion of Natural Function

A. The Problem of the Free Riders

I will now note another complicating factor with regard to the notion of a natural function. We understand that a particular function cannot be the sole cause of the survival of a species (or a particular population). A complex web of factors make it possible for a population to survive given a specific environment. In addition to such factors as undoubtedly contribute to the survival of the population, we may find properties of the body or mind that also survive, partly because they are causally connected to the crucial factors, but that do not contribute positively at all to survival. These factors are, in a sense, free riders in the evolution. (For a similar observation, see Stephen Jay Gould, 2000. He there talks about significant evolutionary features that do not arise as adaptations. He says: "Since organisms are complex and highly integrated entities, any adaptive change must automatically "throw off" a series of structural byproducts ... Such byproducts may later be co-opted for useful purposes, but they didn't arise as adaptations," p. 95).

We can confidently assume that some evolutionary free riders exist in our bodies and minds today. We normally assume that this is the case with the human appendix, for instance. The free-riding organ has never contributed to the survival of the organism, and could be completely neutral in this respect. But we can also imagine cases where members of a species have survived in spite of some trait that they have had and still have. Other traits and properties might have compensated for a relative weakness of the trait in question.

The interesting question for the natural function theorist then is: What should we say about the free riders, the non-contributing organs? Do they not have any natural functions? Can we not find any dysfunctions in such organs?

This appears to follow. Consult again the characterization made by Wakefield: "Certain mechanisms are naturally selected because of the beneficial effects that the mechanisms have on the organism's fitness, and those beneficial effects are the natural functions of those mechanisms." The working of the free-riding organ has, by definition, not been selected because of the benefits of its function for the survival of the organism. This working cannot be an instance of a natural function of the organism in question.

What consequences do these answers have for the theory of health? If some organs or faculties do not have any natural functions, does this mean that they cannot have any diseases or disorders? I will here assume that the appendix in the human body does not have a natural function. But continuous processes are going on in every living individual's appendix. Yet, by definition, no dysfunction can exist in the appendix; the appendix cannot have any disease. But what about an infection in the appendix and what about pain in general that comes from the appendix? Do these not constitute blatant instances of disease?

The following reasonable, but I think insufficient, answer can be given to this argument: Most diseases in the free-riding organs have negative consequences for other organs, including evolutionarily selected organs, and for the functioning of these organs. An infection in the appendix, for instance, may cause general disturbance in the functioning of the organism; it may even be lethal. The infection in the appendix can qualify as a disease, even under the dysfunction interpretation. The disease must then, strictly speaking, be located outside the appendix.

I admit that this answer covers several cases of diseases that we normally locate in free-riding organs. But I contest the assertion that it covers all cases. The contrary case in this paradigm is the local disease that causes pain without otherwise disturbing the major parts of the organism. A free-riding organ may be damaged through an accident, causing considerable pain and preventing the subject from going to work. Here, by assumption, no general infection nor any other general physiological change exists in the body.

A tentative counterargument is the following. A general disturbance occurs in the organism. The pain and disability constitute a general disturbance. This reply is, however, a capitulation. If the answer is that pain and disability can qualify as dysfunctions, then it becomes impossible to uphold Wakefield's distinction between the dysfunction component and the harm component of a disease. Pain and disability are the paradigmatic features of harm in the health context. The harm component must be distinguished from the internal dysfunction.

I conclude therefore that the argument from free riders is a forceful argument against the idea that dysfunction (in the evolutionary sense) is a necessary condition of disease or disorder.

B. The Problem of the Dying Species

Consider also the following case: Some zoologists find a new species on the Galapagos Islands. This is a species, they find out, some members of which have managed to survive in this pocket of the earth, given the peculiar circumstances of the Galapagos Islands. However, now civilization and its consequences have caught up with them, and the species is dying. The population is not, in the long run, capable of protecting itself against pollution and its new predators. All experts expect the remaining population to die within a decade or so. However, so far this population has survived, given a combination of properties of its members and the peculiar, relatively stable environment that has existed up until now.

On the causal interpretation of natural function, the members of the dying species have natural functions: the ones that have, in the past, made them survive. But these functions are now virtually killing the remaining individuals; they contribute to the extinction of this species. Should we then conclude that the features that are killing these individuals represent the natural order and thereby the healthy working of these organisms?

Here also we can expect a counterargument from the natural function theorist. A set of natural (or "healthy") functions cannot protect an individual or a population from extinction if a natural catastrophe occurs. The circumstances can be so harsh that (intuitively speaking) healthy individuals also have to succumb. We can therefore not use a catastrophe as an argument against an evolutionary theory of health.

I grant this argument for the case of the real catastrophe, namely, a sudden event, for example the falling of a meteor to the surface of the earth, that destroys and kills members of a population. My example of the dying species on the Galapagos Islands, however, does not represent such a catastrophe. Instead, it is intended to refer to a rather slow process, over at least a thousand years, during which a population has not been able (or had the time) to adapt itself. This kind of process is not a sudden catastrophe like the falling meteor; it is instead a gradual change of the standard circumstances of the earth; namely, such circumstances as every living being will have to cope with in the foreseeable future. These are the circumstances, to return to the human case, that human beings take as their norm when they subjectively evaluate their health.

5. On the Notion of Exaptation: A Further Problematic Issue in the Theory of Disease as Dysfunction

In some seminal articles, Stephen Jay Gould and Elisabeth S. Vrba (for instance, 1998) have introduced a useful notion into the theory of evolutionary biology: exaptation. This concept is of vital importance for explicating the issues discussed above.

The traditional notion used in evolutionary theory is adaptation. In Gould and Vrba's formulation, adaptation is "any feature that promotes fitness and was built by selection for its current role" (1998, p. 522). This is the criterion of historical genesis as formulated above. But this notion, as we have discovered, does not cover the noteworthy case where a trait, not naturally selected, happens to be useful or can even be consciously developed to be useful for new purposes. Such a trait can, according to Gould and Vrba, be said to be exapted for this new purpose. (To distinguish between the work of an adapted trait and that of an exapted trait, Gould and Vrba use the terms "function" and "effect.")

Gould and Vrba make some valuable points about exaptations. Current uses of a mechanism are often assumed to be the original functions that shaped the mechanism when they are, properly, exaptations. This has led to errors in the reconstruction of the evolutionary history. Gould and Vrba argue, for instance, that the feathers of birds had other functions than flying in the earliest days. More plausibly, they claim, the feathers existed initially for the sake of insulation and protection. It is also probable, they say, that the initial enlargement of the feathers on the narrow extremities first enabled the birds to catch insects. Only later were the feathers exapted for the first flights. Thereafter, these traits can have been adapted in the proper evolutionary sense: They have been selected in the surviving bird populations.

Many uses of mental mechanisms or brain mechanisms in human beings appear to be exaptations instead of adaptations. For simple historical reasons, it is impossible that the functions of reading, writing, and calculating should have been causally selected. Human beings have not, as far as we know, been reading, writing, and calculating for more than roughly 10,000 years. These benefits cannot, therefore, have had any influence on the shaping of the human organism. Reading, writing, and calculating cannot be natural functions in Wakefield's sense.

Only the original function that initially caused the shaping of a mechanism provides an explanation of the mechanism in terms of natural selection. Reading and writing cannot have been original functions of the human brain or mind. They cannot contribute to the explanation of the present human brain or mind. Therefore, they are not adaptations, but exaptations. They are useful applications of the original functions of the brain and mind.

Wakefield has read one of Gould's contributions on this matter and comments on it in his text, "Evolutionary Versus Prototype Analyses of the Concept of Disorder " (1999a). He accepts the importance of Gould's observations, but does not see that these affect his basic analysis of natural function for the purpose of analyzing mental disorder. He says:

> Despite Gould's claims to the contrary, fitness enhancing biological exaptations do require natural selection for their explanation, for two reasons: (a) exaptations almost always involve structural changes that enable the preexisting mechanism, designed for another function, to perform the new function and these changes require explanation by natural selection, and (b) even in rare cases in which exaptations involve no structural change, selective pressures must be invoked to explain why the mechanism is maintained in the population (p. 380).

Wakefield is aware of the problem, but says that it does not affect his analysis because an exaptation can become an adaptation when the trait is maintained in further generations. Gould does not deny this possibility. In the case of birds, it is highly plausible that their capability of flying, realized by the form of their wings, is now an evolutionarily selected function.

But saying that some exaptations can become adaptations over time is not to say that all exaptations become adaptations over time. Therefore, we could never conclude from the occurrence of an exaptation now that it is also a selected trait. My argument about the brief history of the exaptation of the brain for reading, writing, and calculating also shows that in this case, which has major implications for the theory of mental health, we cannot talk in terms of a later adaptation.

It is plausible to say that many mental effects are exaptations. The cultivated human being's young history is a sufficient proof. We have not been able to read, count, and write for more than some 10,000 years. The exaptation of the brain for these purposes cannot have had any evolutionary effects. Time has not been sufficient for a selection process to maintain only that part of the population that can be trained to read, count, and write. What has happened is only that a large part of the population has been exapted for new purposes.

This conclusion may be considered a problem for Wakefield's theory with regard to its application to the theory of mental health. Acknowledged disorders such as *acalculia, dyslexia* and *amusia* exist. These disorders directly pertain to the exapted abilities of human beings to count, read, and express themselves. But if disorders exist that are related to exaptations and not to adaptations, then Wakefield's theory of disorders, relying on adaptations and natural functions, appears to be confronted with significant counterexamples in the core field of mental abilities.

Wakefield has an answer to this objection. To solve this puzzle, he says, we must take the whole notion of mental disorder into account. A mental disorder is not merely a dysfunction; it is a harmful dysfunction. A person's inabilities to count, read, and write are harmful effects of some dysfunction, but they are not in themselves dysfunctions The dysfunctions lie on a lower level of the organism, presumably somewhere in the brain. The proper brain functions must, then, have evolved and be explained by some other benefit for the organism. Wakefield suggests that the *corpus callosum* may fail to perform its natural function of transferring information between brain hemispheres, limiting the intensive information integration required for reading, writing, and counting.

This looks like a winning general strategy to be used for the case of mental health. Most mental disorders can be described as harmful effects of underlying physical dysfunctions.

The harm condition of disorder relates the dysfunction to today's reality. This enables Wakefield to combine the ideas of the historical causal process and of fitness in the present context (although fitness is here not formulated in terms of survival only). A healthy organ must therefore fulfill the historical criterion: It must have been selected for and in a sense also be fit for purposes today. The organ must not create harm today.

How effective is Wakefield's argument? He appears always to have a way out. If something exists, for instance a mental ability, that is considered to be a valuable trait in human beings today but which, from an evolutionary point of view, cannot be anything else than an exaptation, then Wakefield sees to it that this case can still be dealt with within his theory of health. The trait is seen as a positive effect of some basic natural function to be found on the organic level. A deficiency with regard to the mental trait can then still be regarded as a disorder, given the combined theory that requires the existence of a dysfunction in the evolutionary sense and some harm caused by the dysfunction in question.

But such a move does not impress; the theory then becomes irrefutable. Assume that we think we find a counterexample to Wakefield's theory. We find a typical exaptation and we agree that defects with regard to this exaptation are generally regarded as disorders or diseases. Wakefield can then always answer: The defect in the exaptation is only a harm caused by some underlying dysfunction in the evolutionary sense; the exaptation should not in itself be considered to be a function.

But we cannot let Wakefield always escape using this simple formula. It leaves his theory completely untestable. Even if Wakefield cannot tell exactly what is the dysfunction in a particular case, we must at least require that he indicate on which level of abstraction we will identify his functions and that he design a test by which we can find out whether these functions can qualify as natural functions in his sense.

Let this list of problems suffice to question the viability of the evolutionary notion of a natural function for the purpose of clarifying the notion of disease or disorder in general, and the notion of mental disorder in particular.

6. A Sketch for a Different Theory of Health and Illness

The philosophy of health and illness is not helped, I hope to have shown, by a notion of natural function that is based on the historical development of our functions. A much more plausible candidate for this purpose is the idea of a natural function developed by Christopher Boorse. This idea is more plausible since it exclusively refers to the present context and the individual organ's "natural" (as statistically defined) contribution to the survival of the individual and the species today.

Boorse's latest definition of health (1997) is comprised of the following points:

(a) The reference class is a natural class of organisms of uniform functional design; namely, an age group of a sex of a species.
(b) A normal function of a part or process within members of the reference class is a statistically typical contribution by it to their individual survival and reproduction.
(c) A disease is a type of internal state that is an impairment of normal functional ability, a reduction of one or more functional abilities below typical efficiency, or a limitation on functional ability caused by environmental agents.
(d) Health is the absence of disease (1997, pp. 7–8).

The idea of biological function contained here is based on G. Sommerhoff's analysis in his celebrated work, *Analytical Biology* (1950). According to this analysis, a function is a causal contribution to a goal. "An organism or its part is directed to a biological goal G when disposed, throughout a range of environmental variation, to modify its behavior in the way required for G" (Boorse, 1997, p. 9). Since physiology is the subfield on which somatic medicine relies, says Boorse, medical functional normality is relative to the goals that the physiologists assume, namely, individual survival and reproduction.

For reasons developed elsewhere (1995, 2001), I do not, however, find this approach sufficient either. I will briefly rehearse my reasoning. I argue (most clearly in 2001) in favor of the so-called "reverse view" of health and illness (the term has been coined by K. W. M Fulford, 1989). Fulford and I are trying to make a plausible reconstruction of the conceptual field of health and illness in light of medical practice and in light of the concepts of illness and disease as they are understood at present. Here follows a schematic presentation of the steps of development, as we see them.

(1) In the beginning of human existence, people experienced problems in and with themselves. They felt pain and found themselves unable to do what they normally did. They experienced illnesses that were located somewhere in their bodies or minds. Several more people came to experience similar illnesses. This led to the giving of names to the illnesses, whereby their presence could be efficiently communicated. This was, then, the phase of illness recognition.
(2) The persons complained and sought help for their problems. They approached experts on illnesses and asked for cure or alleviation. They communicated with the help of a common illness vocabulary. In the absence of a technical vocabulary, they tried to describe their predicaments wholly in ordinary language terms. This was, then, the phase of illness communication.
(3) The experts—call them doctors—tried to understand the situation. They wanted to help. They tried with the aid of the illness description to achieve a deeper diagnosis. They wanted to have an explanation of the illness. They looked for states and processes that could plausibly cause people's illnesses. Human society approached the phase of disease recognition. Diseases—some internal processes—were viewed as the causes of illnesses, and the effective cures were to be directed at the basic diseases. Illness is, they thought, to be eradicated through the elimination of such diseases.

This is the schematic story that I wish to present as more plausible than its rivals. The most implausible rival story says that a group of biologists started looking for and registered all natural dysfunctions, in the sense of all reduced (statistically determined) functionings in relation to the physiological goals of the survival of the individual and the species, or that they started looking for all imaginable dysfunctions in relation to evolutionary development. According to this story, they thereafter coined terms for all such abnormalities or dysfunctions and called them diseases. Thereafter, they discovered that some of these abnormalities and dysfunctions tended to cause pain, discomfort, and disability. Thereafter, they coined terms for these negative experiences and called them illnesses.

What conclusions can be drawn from Fulford's and my own story? If diseases are potential causes of illnesses, illness is the logically prior concept. The concept of illness determines the concept of disease. The concept of disease is not neutral as to its potential consequences. It instead depends on diseases issuing in illness or ill-being. Or, in positive terms, no biomedical ideal can be construed, this theory says, that is disconnected from human well-being. (This idea of a reverse order between the notions of disease and illness can be traced back to René Leriche and Georges Canguilhem. See Canguilhem's *On the Normal and the Pathological*, 1978, pp. 46–53.)

Jerome Wakefield is not as simple a target as Christopher Boorse is from the point of view of Canguilhem's argument. Since Wakefield has a double criterion of disorder—both harm and dysfunction—he might say that we should first identify the harm; that is, use the Fulford/Nordenfelt criterion. Thereafter, we should distinguish between such harms as are effects of dysfunctions and such as are effects of other bodily or mental states. The first are the proper diseases; the second are not diseases.

The question is: What is the point of looking for dysfunctions in the evolutionary sense, which is a thankless task anyway (as our arguments have indicated)? Would an individual not want to be helped with a problem that has a salient biological or psychological cause, even if neither this problem nor its cause proves to constitute a dysfunction from an evolutionary point of view? And if he or she also legitimately wants to be helped with this problem, and the medical community has the resources to offer such help, then what is the point of establishing that the problem or one of its causes can qualify as a dysfunction in the evolutionary sense?

7. Conclusions

Taking my starting point in the contemporary debate about the notion of mental disorder, I have tried to analyze the notion of a natural function as introduced and presented by Jerome Wakefield. Wakefield proposes a definition of the notion of natural function in evolutionary terms as the only viable notion of function for the characterization of disorder, in the somatic and mental areas.

In my present critical analysis, I have found that the evolutionary concept of natural function is plagued with difficulties. I have raised four main arguments. The argument concerning the degree of a natural function relies on the variability of environments. What was a proper pulse of the human heart a million years ago may not—intuitively speaking—be a proper degree today, but one of the old appropriate degrees must determine the degree of the natural function. The argument from free riders says that several organs are present in existing organisms today that have never been selected. According to the definition of natural function, such organs cannot have natural functions. They cannot be disordered. The argument from the dying species says that the members of a dying population must have some natural functions. Since the species has survived so far—for instance, for a hundred million years—some causes must exist for this survival, in terms of the working of the organs of the members of the species. But in the present environment, these same functions are virtually killing the population. It appears odd to base a theory of health on such functions.

The argument from exaptation says that several human functions (as intuitively understood) have never been selected in the evolutionary sense. This holds true, in particular, for some of our advanced mental functions, such as reading, writing, and calculating. We normally look upon a disability with regard to such functions as a disorder or a disease. However, since these are failures with regard to exaptations and not with regard to natural functions, they do not fulfill Wakefield's criterion of diseasehood. Some failures (with regard to exaptations) constitute criteria for officially recognized illnesses, such as *dyslexia*, *acalculia*, and *amusia*.

Finally, I have sketched an alternative theory of health that does not rely on the notion of natural function at all. According to this theory, the logically primary notion is the notion of illness (entailing harm), from which derives the notion of disease or disorder. The normative notion of harm or one of its specifications, such as suffering or disability, is the crucial notion in the theory of health and illness. But a proper defense of such a theory requires a well-developed argument of its own. (For some developments in this direction, see K. W. M. Fulford (1989) and Lennart Nordenfelt (2001).)

Part Two

CONCEPTS AND MEASUREMENTS OF MENTAL HEALTH. THE CONTEXT OF PSYCHOTHERAPEUTIC TREATMENT

Four

HEALTH, PSYCHOPATHOLOGY, AND THE ACTION OF THE "TALKING CURES"

Einar Jakobsson

1. Introduction

The nature of psychological suffering and adequate modes of psychotherapeutic intervention is a topic of never-ending theoretical dispute. The metaphorical use of the medical discourse in psychotherapy and mental-health work—think of the pervasive use of medical terms like "therapy," "treatment," "clinical," "pathology," "diagnosis," "symptom," "patient," and "illness"—significantly contributes to the ambiguity in this already obscure conceptual domain. In addition to the questionable tendency to "pathologize" human problems, this trend may also lead to difficulties for empirical research in deciding what kind of outcome most correctly reflects the proceedings within consulting rooms.

The aim of this article is to discuss the therapeutic goals and performance of the psychodynamic therapies in relation to their inherent concepts of health. The central issue concerns what kind of influence we can expect when we use psychological means, predominantly relying on dialogue and self-reflection in a dyadic relationship. Should the intended improvement through psychotherapeutic relations be described and evaluated in terms of increasing health, or in terms of decreasing pathology? Further questions result: What difference does it make, for the researcher and for the clinician, to refer to one or the other way of reasoning? What are the possible ways in which these fundamental conceptual orientations may be confounded? And to what extent could an increase in positive mental health be empirically linked to the mitigation of psychiatric disorders and symptoms?

My thesis is that we should evaluate the modern relation-focused psychotherapies of psychodynamic origin primarily, though not exclusively, as health-promoting activities, rather than interventions directed at the cure or amelioration of symptoms or psychiatric disorders. Most likely, my conclusions will be relevant to many similar "talking cures" of experiential, systemic, or cognitive origin, but strictly speaking, they apply only to those psychotherapies that emphasize the dynamics of the therapeutic relationship as a significant means to understand and help the client. We cannot clarify the action of these therapies, distinguished as they are by their universalistic means for the benefit of holistic and heterogeneous goals, by ruling out the possibility of placebo or spontaneous remission in accordance with the medical model and the efforts to compare the outcome of one measure with another on some pathological condition.

2. Psychotherapy Research

According to Klaus Grawe and Mariann Grawe-Gerber (1999) and many other researchers in psychotherapy, the psychotherapeutic task is best described and evaluated in terms of its resource-activating power. Such an approach indicates that we should focus on the ability of psychotherapeutic interventions to enable the client to make use of locked-up potential and to cope with his or her life problems. Using Aaron Antonovsky's formulation of health in terms of sense of coherence, we could say that this task might have more to do with salutogenic (health-promoting) factors than the elimination of pathogenic factors (Antonovsky, 1987).

Furthermore, a widespread conclusion of modern psychotherapy research that is in line with that view is that we can seldom show the effects of one type of psychotherapy to be superior to another. Consequently, many researchers during the last two decades have turned their attention to issues regarding the therapist and/or the therapeutic relationship, factors in medical thinking that are usually considered as having placebo or non-specific influence. A considerable part of recent research concentrates on aspects of the therapeutic process—that is, the therapeutic alliance (Horvath, 2000) or the detection of significant events (Elliott and James, 1989; Elliott and Shapiro, 1992)—instead of outcome in terms of a main effect on some diagnostic category.

At the same time, another part of continuing research still centers on the question, "What works for whom?" Such studies understand this kind of inquiry as a matter of which type of psychotherapy or technique is working for what kind of pathological condition (see, among others, Roth and Fonagy, 1996). Such a specification of the effects of psychotherapeutic means may be relevant, though difficult to bring about in the case of most "talking cures." Psychodynamic therapies do not usually cling to detailed manuals that easily lend themselves to operational definitions and randomized, controlled trials. Roth and Fonagy admit that psychodynamic therapists transcend narrow diagnostic categories in their work, and claim that "the particular value of psychodynamic formulations is the degree to which they locate the present functioning of a patient in the context of the individual's life experiences" (Roth and Fonagy, 1996, p. 44). If we accept this premise, then relief from psychiatric symptoms associated with some pathological conditions should not be the only relevant assessment.

Some psychodynamic psychotherapy researchers do make use of operational definitions—even of transference, the key concept of psychodynamic therapy—and thus exemplify a new striving within a tradition that hitherto has mainly confined itself to case studies or evaluations based on a limited number of self-reports. But in this case, the efforts toward specification concern possible alteration of the internal representations of self and others, not the effects on diagnostic categories of mental illness (Luborsky and Crits-Cristoph, 1998).

3. Notions of Mental Health in Psychodynamic Psychotherapy

In the brief exposition that follows, I will also compare the holistic theory of health espoused by Lennart Nordenfelt (1995a) to the notions of psychotherapeutic health embedded in psychoanalytic theories of development, psychopathology, and psychotherapeutic processes. The question is to what extent they may be compatible, and if this comparison may shed some light on what modern psychodynamic psychotherapy is accomplishing.

According to Nordenfelt, health in general should be considered not just as absence of disease, but as something standing by itself to which pathology or diseases relate as only one set of compromisers among others. This action-theoretic definition of health states that a person is in health if he or she has the ability to reach his or her vital goals given standard or accepted circumstances. Ill health, not disease, becomes the opposite of health, and a person with a disease can be relatively healthy or be in full health as long as he or she is able to reach all of his or her vital goals.

Why not, then, concentrate on explicit psychodynamic or general formulations of mental health? Only a few psychoanalytic writers have paid substantial interest to conceptual issues regarding the concept of mental health in relation to the psychotherapeutic task. The notions of health in psychotherapy theory are usually implicitly equated with the desired psychodynamic changes in (aspects of) the personality and conceptualized at an "experience-distant" high level of abstraction—and, usually, in connection to a model of psychopathology. Put another way, theorists commonly conceive of mental health as tantamount to the general aims of psychotherapy, although these might be viewed both as absence of psychopathology and as enhancement of some functional aspects of the personality, notably a sense of mastery and satisfaction in human relations.

Marie Jahoda's (1958) review of psychological health formulations is still current and often cited. Her suggestion of a comprehensive and positively defined concept of mental health relies on multiple criteria and points to well-known aims and values of the talking cures—like "attitudes toward the self," "growth, development, and self actualization," "integration," "reality orientation," "autonomy," "perception of reality" and "environmental mastery." Psychotherapy theory includes many similar concepts.

I have elsewhere outlined the multiplicity of these formulations under four fundamental headings: knowledge, autonomy, adaptation, and well-being (Jakobsson, 1994). We may see knowledge (or self-understanding in a broader sense) here as a goal/value in itself or as a sub-goal or means to value-laden concepts linked to the other three headings. Conditions like maturity, ego strength, and integration (linked to the qualities of adaptation) and self-esteem, vitality, and sense of coherence (linked to the qualities of well-being) would then be prerequisites for the individual's ultimate goal of self-realization. Most of these positive formulations refer to an abstract, theory-laden level that at the same time points to pathogenic contingencies that may block optimal emotional development and the prospect of self-realization.

4. The Issue of Underlying Psychopathology and Mental Health Goals

Quotations from three influential psychoanalysts who have played a key role in modern psychodynamic psychotherapy provide an illustration of the general psychoanalytic tendency to reduce human psychological suffering to some crucial pathological conditions that are attributed a causative role. Melanie Klein's theoretical contribution to psychoanalysis is based on theories of early experiences of love and hatred and their management in relation to the mother. Her definition of mental health uses the following well-recognized concepts:

> A well-integrated personality is the foundation of mental health. I shall begin by enumerating a few elements of an integrated personality: emotional maturity, strength of character, capacity to deal with conflicting emotions, a balance between internal life and adaptation to reality, and a successful welding into a whole of the different parts of the personality (Klein, 1960/1975, p. 268).

While psychopathology in Kleinian theory mainly concerns conflicts that feature greed, hatred, envy, and rage, which are seen as the concomitants of constitutionally based aggression, the reparative and integrative activities in connection with the likewise inherent life instincts (love and gratitude) should be activated in the healthy individual. Psychotherapy and good mothering could thus be considered as health promotion that aims to balance pathological or pathogenic elements of the personality and make possible an integration of good and bad, where the good is experienced as predominant. A person in health could then be characterized by his or her emotional maturity, manifested as psychic strength to deal with inner conflicts and a mental ability to adapt to reality.

Otto Kernberg's psychoanalytical writings reflect a traditional medical orientation with a steady focus on diagnosis of psychological disorders and the therapeutic means for the solution of the underlying psychopathology. Talking about narcissistic patients, he claims:

> However, whenever possible, one should attempt to resolve this serious psychopathology in order to achieve a fundamental change in the patient's internal relationship with himself and with others—a shift, in short, from pathological narcissism to normal narcissism and object relationships (Kernberg, 1975, p. 293)

From Kernberg's point of view, the elimination of the underlying pathology is obviously necessary for the restoration of normality and health in the sense of an increasing capacity for establishing deep and enduring relationships with others. In this he is in agreement with the classical Freudian attitude, which emphasized insight into the contextual origins of the psychological problems.

Heinz Kohut, the creator of a psychoanalytic psychology of the self, is one of the few psychoanalysts who have discussed psychotherapeutic "cure" in relation to mental health. He passes criticism on the tendency of his forerunners to associate cure (or full mental health) with some solution to a crucial psychological step in early development in a way that illustrates the topical issue:

> The healthy self that may ultimately establish itself may deviate from norms derived from the assumptions that every analytic cure is defined either—according to the traditional Freudian outlook—by the engagement and ultimate resolution of oedipal conflicts or—according to the Kleinian view—by the reexperiencing and ultimate overcoming of the most archaic layers of depression, suspicion, and rage (Kohut, 1984, p. 7).

Kohut himself, however, does not differ from other psychoanalysts insofar as he primarily defines the healthy (and the unhealthy) person in metapsychological terms with reference to the importance of some line of (self)development in the past. Mental health becomes a matter of the structural composition of the self, that is, the reactivation of compensatory structures, which enables the self to seek out and use "empathic resonance" for its own sustenance (Kohut, ibid., p. 77). When this is the case, the unique combinations of ideals and talents rooted in a nuclear program at the core of each individual self will become realizable and lead to an inner completeness experienced as a sense of coherence, vitality, and a meaningful life.

The therapeutic method is interpretation as usual, but with a special emphasis on the corrective emotional experience it will bring about through empathy. The effect of this kind of intervention is called transmuting internalization, meaning that it is an optimally frustrating explanation, empathetically conveyed by the therapist, that will correct and cure deficiencies of the self that once resulted from lack of responsive care or traumatic experiences of childhood. Such experiences are then seen as the crucial pathogenic condition from which most psychological problems can be derived.

The therapy employed will vary with the theoretical persuasion held by the psychotherapist and with the different problems and needs presented by the clients. The psychodynamic psychotherapist, however, is likely to subscribe to some general ideals of personality (for example, psychic strength, social competence, and solid identity), which may or may not be extended to some ideals in life (for example, self-realization and a satisfying/meaningful life). These ideals usually connote the therapist's notion of full mental health, and consequently these inevitably heterogeneous goals are, as Alan C. Tjeltveit (1999) also argues, inextricably linked to the ethical ideals of the psychotherapy/psychotherapist.

Two types of goal theories dominate the psychotherapeutic contribution to the notions of mental health: (1) those that emphasize the crucial role of some kind of pathology for ill health, and (2) those that emphasize the optimal development (or therapeutic strengthening) of some value-laden mental capacities that are associated with ethical considerations of a client's satisfactory functioning.

Is this sufficient evidence for the conclusion that these theories latently also correspond to a holistic health concept like Nordenfelt's? Mental health is considered as compromised by a pathology that must be resolved or mitigated in order to make it possible for the client to reach his or her vital goals. These therapeutic goals include changes in the client's mental abilities, such as adaptive capacities, that may or may not be deemed as vital in themselves or vital for reaching other goals (like well-being or fulfillment through some maturational process). If we use a formulation borrowed from Tjeltveit (1999, p. 201), we generally have difficulty deciding whether what is considered as essential for the psychotherapeutic task is the freedom from psychological problems (psychopathology) or freedom for some more-or-less optimal state of being called mental health.

The distinction between freedom from and freedom for also depends on how pervasively "ill" the client is, and whether this "illness" is exactly what the psychotherapist is expected to treat. The difference between psychotherapy with a client suffering from psychosis and psychotherapy with a client complaining about anxiety or life dissatisfaction is significant. The goal of the intervention, however, may in both cases be a matter of enhancing the client's psychic strength and coping ability, instead of changing the pathological condition or symptoms connected to the mental disorder itself. The kinds of complaints or problems presented are as essential as the client considers them to be for the realization of vital goals.

Sometimes a psychotherapeutic goal may coincide with the realization of a client's life goal, rather than serving only as a means for reaching some extra-therapeutic goal. Regarding Nordenfelt's view, his definition maintains that fulfillment of vital goals is not demanded of an individual in health; the ability to realize them is sufficient. Therapeutic goals that aim at freedom for something usually imply a restoration or development of some capability that at the same time materializes in terms of its valued consequences.

5. Rhetorical Means and Mental Health

Irrespective of these varying views on the ultimate reasons for psychological suffering, and irrespective of how much attention a therapist pays to the resolution of some kind of psychopathology, the central means of the psychodynamic therapies has always been interpretation. As with insight (and many other concepts in psychoanalytic theory), the meaning of interpretation is altered in accordance with the theoretical context. I use the term here in its most extensive sense to denote all of the psychotherapist's rhetorical activities that reflect his or her efforts to understand and explain the client's situation. Interpretation thus includes comprehensive constructions and limited comments, within the transference or outside it, referring to the ongoing therapy relation or to other (former and contemporary) relationships of the client.

The action of the therapy relation as a whole on the internal world of the client—or on the affect-laden representations of self and others that may be seen as the basic constituents of a *Weltanschauung*—may derive from several factors that are related to the therapist's handling of the therapeutic dialogue. The principal factors in that dynamic are: (1) suggestions, for instance, the induction of hope and safety in conditions of demoralization; (2) catharsis, or focusing on expression and re-experience of emotional trauma or distress; (3) empathy, including both acceptance and understanding; (4) corrective emotional experience, or the client's identification with the therapist, some therapeutic attitudes, or the therapeutic relationship as a whole; and (5) insight or, more generally, integrating self-knowledge, including identity strengthened through narrative reconstruction of the client's internal representational world, past and present. These factors should not be considered as excluding one another, but as coexisting in different combinations. A reasonable supposition is that the last factor (5) usually integrates the ingredients of action into a complex effect on the client's outlook on life and psychological functioning.

Modern psychoanalytic thinking has acknowledged this effect as a shift of emphasis in psychotherapy theory, a move from interpretation (in the restricted sense of uncovering repressed memories and conflicts) to relation and analytic autobiographies (Meissner, 1991; Blum, 1992; Sandler and Dreher, 1996).

The concept of action is also relevant for the distinction between a problem- or pathology-oriented and a resource- or health-oriented approach to the intervention. The alternatives could emerge in the focus in the psychotherapeutic process—which may be on efforts to understand and work through the underlying pathological obstacles to emotional development, or on efforts to activate mental resources (for instance, by the narrative reinterpretation or molding of the experience of the self in relation to others). Yet I expect that a vast majority of contemporary psychodynamic therapists would say that they operate with both foci in mind.

As to the question of action, a majority would also admit that cognitive insight into unconscious conflicts and defenses (or good and bad representations of self and others, or deficits of the self) cannot by themselves explain the effects of psychotherapy. Some of those qualities of the therapy relation—for instance, the internalization of the conduct and analytic attitude of the therapist—may be just as significant in a complex process that contains more than one type of influence from the therapy relation experience itself.

If we dared to make a synthesis out of these possible action mechanisms, the product would be a mixed picture of an experience, from the client's perspective, of emotional support and cognitive orientation/understanding that resulted in an increase in psychotherapeutic mental health. The psychodynamic therapist thus helps the client

(1) to express his or her complaints and psychological suffering
(2) to express and explore his or her internalization of the affective representations of the world (self and others), both past and present
(3) to express and explore his or her experience of the therapy relationship
(4) to make a narrative reconstruction of his or her personal life out of the elements of (1), (2), and (3).

The theories of underlying psychopathological development, which often is seen as the chief compromising factor to good health, play the role of "illness" or even "disease." The emphasis the individual therapist places on pathology, in relation to the elements mentioned above, determines the relative importance of an analysis and mitigation of these blocking elements, which consist of dysfunctional affective beliefs that are part of the client's internal belief systems.

My characterization of the talking cure, then, is in agreement with Stephen Toulmin's (1978). He describes such a cure as an activity aiming at remedies to complaints (rather than cure of diseases) and, hence, an activity with its own virtues. The prime virtue would be self-reflection or self-realization that might lead to a new mode of life. These virtues are transmitted by the rhetorical reinterpretation of selected parts of the client's internal world, which are deemed as vital for the increase of his or her health. The ancient Greeks, as Robert Spillane (1987) notes, were already convinced that "physicians of the soul" are concerned with rhetoric rather than medicine, conversation rather than cures.

6. Conclusions

My first conclusion is that the general good that the psychodynamic psychotherapist aims at is an increase of the client's self-knowledge to the benefit of his/her mental health. This health-promoting self-knowledge should primarily manifest itself as an improvement of the capacity of self-mastery in human relations. This goal can be accomplished through rhetorical (interpretative) activities in a secure setting that offers a therapy relation experience built on understanding, hope, and safety. The content of the interpretations may vary and, hence, the insight or the general self-knowledge resulting from it. However, an essential reference to some therapy relation-based teachings about the constituents of a healthy (as well as unhealthy) personality and the good life is always present.

In other words, interpretation may in modern psychodynamic therapy aim at everything from Freudian uncovering of unconscious conflicts over the enactment of the internal world of object relations to the narrative retelling of lives and belief systems based on social construction theory (Stern, 1996). The acquisition of self-knowledge may be seen as an essential life-goal in itself or as a means for reaching further goals of self-realization, the achievement of which may be noted as a sign of optimal emotional maturity or full mental health.

My second conclusion is that psychodynamic thinking is irrevocably interwoven with theories of developmental core pathology. But when we ask what the therapist can do about this, the general means and the holistic, heterogeneous goals of the therapeutic intervention indicate that the action basically relies on the "common curative factors" suggested by psychotherapy research. A reasonable hypothesis is that psychodynamic change of (aspects of) the client's personality is a matter of activating, strengthening, balancing, and correcting mental resources and representations in favor of the client's adapting or coping ability and sense of well-being.

This change may have secondary, mitigating effects on different symptoms and disorders or on the ability to cope with them through the new, affect-laden representations of self and others. Insofar as the symptomatic condition or disorder is linked to its symbolic representations, we could expect a radical change of symptoms together with the underlying "pathology." Getting rid of the problem then becomes a matter of dealing with it in terms of how it could be reinterpreted and mastered.

My third conclusion is that the concept of mental health in psychodynamic psychotherapy is related to Nordenfelt's concept of health in interesting ways. Nordenfelt states that a person is in health if he or she has the ability to reach his or her vital goals given acceptable circumstances. Psychotherapeutic help offers a client the chance to change (aspects of) his/her personal functioning in order to be able to realize goals that may be deemed vital to his or her participation in the sphere of human relations. If the therapy is successful, shouldn't we then conclude that the client has increased his or her health due to some kind of self-improvement that includes an increase of goal-related capacities?

The key concepts are ability and life goals, and Per-Anders Tengland's (1998) analysis of mental abilities necessary for having "acceptable mental health" is relevant here. He emphasizes some basic abilities concerning communication, empathy, flexibility, perception, and problem solving—in short, the ability to cooperate and the ability to have practical rationality—as vital for survival in modern societies. From a psychotherapeutic point of view, I think that these basic abilities usually are regarded as prerequisites of the psychotherapeutic intervention. In psychotherapy, therapist and client have in mind the refinement of these abilities and the prospects of full (or optimal) health.

We have already seen that the therapeutic goals in psychodynamic thinking are heterogeneous and may encompass the realization of life goals. However, psychoanalytic thinking sometimes differentiates between therapy goals and life goals. This equates roughly with the aforementioned distinction between freedom from something bad (for instance a pathological condition) and freedom for something good (for instance, self-realization), between interpretations of symptomatic behavior and interpretations of life-worlds.

Some psychodynamic therapists would, like Sigmund Freud, confine themselves to making the client "fit for existence" (Freud, 1905/1960); the holistic approach aims at understanding the blocking elements in their contextual origin and range. The resolution of pathology and/or the strengthening of the abilities to cope with your life situation would then be a therapy goal, while the use of adaptive abilities for self-realization is left to activities outside the therapy process.

Others would, like Kohut, search for meaningful content in life and try to establish a sense of coherence and well-being according to the needs of an individual client. Both attitudes, however, emanate from the imperative for some kind of self-knowledge. This self-knowledge is a value or virtue in itself or, in the case of upholding a distinction between therapy goals and life goals, a means for other virtues as well. Both attitudes also estimate—in a self-referential way common in psychoanalytic theory—the goals of psychotherapy, and the experience of therapeutic relationships, as vital for a client's health.

Comparing this reasoning to Nordenfelt's concept of health, we can finally conclude: first, that psychodynamic literature has a holistic approach and provides occasional comments implying that mental health is something more than the absence of "disease." However, psychodynamic theory, as a rule, does not regard health as something standing by itself. The reason is that psychoanalytic/psychodynamic theory, being an imposing theory comprised of many theoretical aspects, always contains a theory of psychopathology that is meaningful for psychotherapeutic practice and accompanying notions of health. At the same time, with respect to recent psychotherapy theory, it would be possible for contemporary psychodynamic therapists to subscribe to a free-standing concept of health. This can be done with special reference to modern (as well as ancient) theories of resource-activation, sense of coherence, and rhetorical action.

We may also conclude that psychodynamic theories tend to be content-oriented in a way that goes beyond Nordenfelt's formulation in principle. This means that psychodynamic notions of health have become increasingly oriented toward philosophies of life connected to pragmatic solutions to problems in human relations. The ability to reach vital life and health goals in this area is therefore linked to a set of concepts that are related not only to general principles about psychic strength and adaptive personality functioning, but to ethical versions of the human condition and the good life.

Five

A CONCEPTUAL APPROACH TO MEASURING PSYCHIATRIC AND PSYCHOTHERAPEUTIC TREATMENT

Per-Anders Tengland

1. Introduction: What Do Psychiatric and Psychotherapeutic Instruments Measure?

Psychiatric and psychotherapeutic research literature contains several instruments, mainly quantitative, that measure the outcome of psychotherapeutic and psychiatric treatment (Bergin and Garfield, 1994; Sandahl, Birgerson, Armelius, Fridell, and Segraeus, 1998). We also have plenty of screening instruments that measure the mental status of the population (McDowell and Newell, 1996). But what are these instruments supposed to measure? These tools ask similar questions, but the names of the instruments vary (McDowell and Newell, 1996, p. 122, p. 178). Do they measure mental health, well-being/suffering, personality, emotional adjustment, quality of life, performance, social functioning, mental illness, disturbance, or disorder? And what exactly do these terms mean? One reason for this obscurity is that the instruments do not always have a clear conceptual basis. This factor affects their validity negatively. We cannot be sure that the instruments measure what they purport to measure.

Instruments are usually constructed through an empirical approach. Many researchers have tried to find criteria for what they want to measure by asking relevant people for their recommendations. This method has its drawbacks. But instead of criticizing such instruments in detail, which I have partly done elsewhere (Tengland, 2001), I will discuss an alternative way to approach the construction of an instrument for measuring outcome or mental-health status. Using this method, the researcher takes a conceptual discussion as the starting point of an assessment. Having a solid conception of what is to be measured makes it easier to choose what questions to ask: The questions chosen are those that are logically and/or causally connected to this conception. This approach helps increase a study's validity.

In this article I intend to show how researchers can go about the task of creating such a measuring instrument. I will be concerned only with the question of measuring mental health (personality, well-being, etc.) in general, not with screening for some specific mental disorder, like depression or dementia.

2. The Goals of Psychotherapy, Psychiatry, and General Health Care

I believe that only two general goals exist for psychotherapy, psychiatry, and health care in general: health and quality of life. All other health-care goals must be related to these goals, either logically or causally. Some authors hold that quality of life, together with a long life, is the only goal that has intrinsic value (Brülde, 1998). Even if this is true, it should not be the primary goal for most health care.

General health, I claim, is the goal for most health-care interventions. Included in this goal is not only an increase in, or the restoration of, health (for those who are in a state of ill health), but also a reduction of the risk of losing health (health promotion and disease prevention). Most other health-care goals can be reduced to this general goal. One exception is palliative care, where the reduction of suffering, even if it does not lead to the increase of health (as the term will be defined), should still count as a genuine goal of health care.

Quality-of-life goals are secondary goals within health care, in the sense that an increase in quality of life (in this context) should be an effect of raising the general health of the individual. We might even claim that an increase in the person's health has to lead to an increase of quality of life to be a proper goal for health care.

Psychotherapy and psychiatry strive for an increase in the mental part of what constitutes health, and since we will be discussing the outcome of psychotherapeutic and psychiatric treatment, we will here primarily be concerned with the mental part of the person. For these claims to be understandable, I need to specify what I mean by health, mental health, and quality of life.

3. The Need for a Theory of Health and Mental Health

Discussions of mental health must start with a theory of general health (Tengland, 2001). The reason is the following: For any statement of the kind "Rationality is a requirement for having mental health," "A mentally healthy person must show some capacity for anxiety tolerance," "A healthy person is a person who can love," or "A mentally healthy person has self-confidence," we can ask, "How is it justified?"

To answer that question we need a touchstone: a concept of general health. Plausible answers to that question in reference to the statements above are: "Rationality is part of mental health because it is part of the human design," "Rationality is part of mental health because it is how people normally are," "Rationality is part of mental health because it promotes survival and reproduction," "Rationality is part of mental health because it furthers our well-being," and "Rationality is part of mental health because we need it in order to be able to reach significant goals in life."

Each of these answers indicates some kind of general theory of health. The level of understanding they reflect is more or less rock bottom; we do not need to ask why it is valuable to survive, have well-being, or reach meaningful goals. As to having intrinsic value, quality of life and longevity may be more fundamental than health, but to some extent these are already built into most conceptions of health. If the mental feature in question leads to survival, well-being, or the attainment of meaningful goals, it thereby also leads to an increase in quality of life or longevity. But quality of life is only implicitly (if at all) built into more biological definitions of health as "normality" or as "evolutionary design."

I can envisage four or more general ways to define positive health: as normality (Boorse, 1975, 1976, 1977); as balance (Pörn, 1984); as general well-being (WHO, 1948); and as general ability (Nordenfelt, 1995a, 2001; Pörn, 1993; Fulford, 1989; Whitbeck, 1981). A fifth possibility is a multidimensional theory (Brülde, 1998, 2000). (Health can also be defined negatively, as the absence of disease.)

The literature contains three major ways to characterize quality of life: as "pleasure" (the hedonistic theory), as an objective list (the objective list theory), and as wish-fulfillment (the equilibrium theory). I will not discuss quality-of-life conceptions in detail, and will here refer only to the third kind of theory, wish-fulfillment. According to this theory, to have quality of life is to have many or most of your principal wishes fulfilled. However, added to this conception is that this fulfillment, were it to be actualized, should lead to the disposition to feel happy. A loose characterization of the position is the following: A person has high quality of life iff (if and only if) most of the principal wishes the person has are fulfilled, and he or she has a tendency to feel happy when reflecting upon this fact (Nordenfelt, 1993a).

4. The Holistic Theory

For the sake of brevity, I will not discuss the pros and cons of the health theories mentioned (I have done so elsewhere: Tengland, 2001). I will, in this article, use Nordenfelt's holistic theory of health (Nordenfelt, 1995a, 2001). In this theory, "ability" is the central term in defining health. The definition of health put forward by Nordenfelt goes like this:

> An individual is healthy iff the individual has the (mental and physical) ability to reach his or her vital goals, given acceptable circumstances.

This theory is holistic, and it is so in two senses. First, it takes the whole person as its starting point, not the parts or functions of the individual. Second, the theory is holistic in that it also takes the individual's goals and his or her environment into consideration.

Two terms in the definition need further elucidation: "vital goals" and "acceptable circumstances." "Vital goals" is a technical term that refers to goals that together, if reached, make the person at least minimally happy. More formally: x is a vital goal for P iff x is a necessary condition for the minimal long-term happiness of P. The term "minimal" indicates that there are other factors in life that influence happiness, factors that do not depend on the individual's own ability. That is, for maximal happiness, other wishes about the state of the world have to be fulfilled.

The individual usually chooses his or her goals in life believing and hoping that the attainment of them leads to happiness. However, the individual can be mistaken about which goals lead to happiness. Conditions (possible goals) unknown to the individual might exist that, if reached, would make him or her happy.

Acceptable circumstances are circumstances under which we can generally expect people to be able to achieve their goals. What exactly these circumstances are is something we have to agree upon. To give an example: a situation like civil war, or some other catastrophe, might stop persons from reaching their goals in life. Under such conditions, we would not claim that these persons are unhealthy if they do not reach their vital goals, since these are not conditions under which we expect people to be able to do so.

5. Suffering

Since ability and disability are the central terms in Nordenfelt's theory, how are mental states like well-being and suffering related to this theory? In choosing which of the terms disability and suffering is the most central in a conception of ill health, Nordenfelt comes to the following conclusion: Disability is more central than suffering because we can find examples of ill health where suffering is absent. A person in a coma is in a state of ill health, but does not experience suffering. The same goes for a person paralyzed from the waist down. However, in both of these cases the person is disabled.

Is it not also possible that there is suffering without disability? This depends on the definition of the terms used. According to Nordenfelt, suffering is a mental state that by definition causes disability. A condition cannot therefore be a state of suffering if it does not causally reduce the individual's general ability. In line with the theory used, this is the conception of suffering that will be used here.

6. Mental Health

What is mental health, according to this theory? Formally put, mental health is the mental part of the general ability to reach vital goals. However, we also want to be able to say something more concrete about mental health, that is, put it in material terms. One suggestion is that we can divide the mental into several mental abilities. Another, more biological, way to describe the mental apparatus would be to speak about functions. But "ability" is a commonsense term that is broader than the term "function," covering complex abilities like the ability to drive a car and the ability to make friends. I therefore find it more suitable for this discussion.

In general, the term "ability" is used for intentional activities, like having the ability to walk, reason, or decide. However, the term will here be used in a broader sense, as also meaning being able to be in a given mental state (having designated dispositions). More formally, the term will be used in this sense: "P has the ability to x iff P has the internal properties that make it possible for x to, or to be in state, x." Thus we can say that an individual has the ability to feel love and compassion, or is unable to feel anger or sadness. This does not imply that emotions or feelings like these are intentionally created.

What specific mental abilities are then needed in order for the person to have mental health? The mental apparatus can crudely be divided into three areas: the cognitive, the conative, and the emotive. Each of these areas consists of abilities; for instance, the ability to remember, the ability to want something, and the ability (or disposition) to experience anger. Some abilities relevant to having health are more complex; for instance, the ability to communicate and the ability to cooperate. Several mental abilities exist that are necessary, to some degree, in order to reach most vital goals (Tengland, 2001, pp. 104–138). Some examples of mental abilities that are likely candidates for being part of the mentally healthy person's makeup are: rationality (epistemological, practical, logical); correct perception of reality; memory (long term, short term); flexibility (in thinking and acting); the ability to communicate; self-esteem; empathy; and volitional power.

7. With Respect to Instruments, What Follows from the Holistic Theory?

Given this theory of health and mental health, we run into a general problem if we are to create quantitative instruments in order to measure health. Since the goals (believed to be vital) of the individuals can be different, and people therefore need to have different abilities to achieve them, it appears that we cannot construct an instrument that is applicable to all people.

Below I formulate five consequences, given the theory used, with regard to constructing an instrument: Consequences A and B are ways to deal with the problem of the relativity of goals, C and D follow from B, and E is a further suggestion indicated by the theory.

A. Qualitative Aspect

One solution to the general problem of individuals having different vital goals is to give the instrument a qualitative part where the vital goals of the individual are explored. This means that it would be necessary to ask each individual explicitly about his or her goals, and whether the ability to attain them has changed during therapy. (An additional problem here is that the goals might have changed during therapy.) I will not, in this article, discuss the implications for qualitative studies.

B. Basic Vital Goals

Another way to address the problem of the relativity of goals exists. People's goals are, on an abstract level, fairly similar. I have suggested that we postulate some fundamental goals, goals that are fairly common to all people. I call them "basic vital goals." These goals are to hold some kind of job, to take care of yourself. your home and your finances, and to have some social contact (Tengland, 2001, pp 97–100).

C. Mental Abilities

Given that we can postulate the basic vital goals under B, we can also ask questions about specific mental abilities needed in order to reach these goals. A partial list of such abilities was presented above. The idea is this: An increase in the abilities needed to reach basic vital goals is an increase in mental health, and thus health, but is most likely also an increase in the ability to reach many other kinds of goals.

D. Mental Suffering and Well-Being

Suffering tends to reduce the general ability to do things—for instance, reaching basic vital goals—usually by reducing specific mental abilities. In the mental realm, we find that mental suffering reduces mental abilities, and that well-being increases these abilities. Thus, we can ask questions about states of mental suffering and well-being, and draw conclusions about mental health from the responses to such questions. Physical pain, no doubt, also reduces mental abilities, but the problem will not count as mental ill health since the cause of this reduction is physical, not mental. In practice, it might sometimes be hard to determine whether a problem is physical or mental, as in psychosomatic problems.

E. Theoretical Criteria

The conceptions of health and mental health here presented are more or less neutral in relation to psychological theories, since they are mainly based on an ordinary, common-sense conception. This conception assumes, however, that mental abilities (including dispositions) exist. These are not "observable," and some psychological theories, for instance behaviorism, might have problems with this characteristic. But on the whole, as these abilities are presented in ordinary language, they should be acceptable to most researchers.

The advantage with an instrument that does not rely on theoretical terms is that it is easier to get researchers from different schools to accept the same results. In addition to the "theory-neutral" results reached, researchers of different orientations can add, if they want, their theoretical interpretation of the results. Evidently, though, the terms are not culturally neutral (Danziger, 1997).

8. Creating an Instrument: Theory

When creating an instrument for measuring mental health (as it is defined here), we can ask three general kinds of questions: questions about mental abilities (refers to C above), questions about general health (refers to B), and questions about mental well-being/suffering (refers to D). The last two categories measure mental health indirectly. Questions about general health are not necessarily about the mental. A positive answer to a question about general health indicates mental as well as physical health, since both are part of the general ability. But a negative answer will not tell us whether the problem is mental or physical. Well-being questions are causally related to mental health, since positive well-being in general affects the mental abilities positively, and suffering affects them negatively.

A. General Ability, Reaching Basic Vital Goals

We can suppose people's goals, on an abstract level, to be fairly similar. I have called these goals "basic vital goals." When creating an instrument, we can ask in general if the individual has the ability to reach these basic vital goals. Questions might also exist about some holistic (mental) abilities that are required for reaching these basic vital goals.

B. Specific Mental Abilities

Given the theory used, this is the most relevant category, since these questions are about mental health. Crudely, the questions can be divided into questions about cognitive, volitional, and emotive abilities. This will include some mental dispositions. The first group is likely to be the largest one; thus, we can expect more questions about this category.

C. Well-Being/Suffering

Moods and emotions play a causal role in health and ill health, since suffering affects the cognitive and volitional abilities negatively, and therefore also more complex abilities like communication and cooperation. Mental suffering is a key category, since it causes mental disability. We can then expect that questions about well-being/suffering can give us useful information about the individual's mental-health status.

Note here that according to the theory used, the inability to experience some emotions—say, love—can be an example of ill health (where this disposition is absent), whereas experiencing anxiety is only causally linked to ill health; for instance, when this anxiety reduces the person's ability to act.

9. Possible Problems

A. Physical or Mental?

One possible problem with the instrument has to do with physical illness, and primarily concerns the first category (the general ability to reach basic vital goals). The problem is that it might be hard to determine whether the person answering the questions has problems reaching these goals because of a physical or of a mental disability. So we also need other questions, about specific mental abilities, thinking, memory, will, emotional capacity, etc., and about mental well-being or suffering.

B. Health or Quality of Life?

We might also have the problem of knowing whether a question is a question about a mental-health issue or about a quality-of-life issue. A question like, "How do you feel about the future?" will probably be answered in the negative, when answered by someone suffering from a mental problem and by someone with a poor quality of life. To minimize this problem, we have to be careful about how we formulate the questions.

We have to remember that health and quality of life are causally connected notions. Poor quality of life, no doubt, increases the risk of reduced health, and ill health is not only part of poor quality of life (since it is not wanted by the individual), but most likely leads to even lower quality of life.

10. Preliminary Questions

Before sketching more concretely what an instrument of this kind can look like, I have to take care of some preliminary questions.

First, we need to formulate the questions taking into account that we want to measure a state that has some duration, and is not just a temporary and transient one. We can find many ways to do so; I have chosen to precede each question with the phrase "Have you lately ...?"

Second, an even more relevant issue exists that must be decided before we can start formulating questions: Should the questions be asked in the negative mode, focusing on ability problems, negative feelings, etc.; in the positive mode; or in both modes?

The advantage with only asking questions that focus on the negative side is that this makes it easier to avoid questions that also indicate quality of life. A suggestion in accordance with this model is that each question ends with a clause of the kind: "... and this is not due to recent negative or tragic external events in your life (like losing a job, going through a divorce, or losing a loved one)." It is much harder to include a general clause like this one if the questions are formulated in the positive mode, and impossible if the questions are mixed. The other solution to this problem is to start the whole questionnaire with a clause that tries to rule out that the questions are about positive or negative external events.

Norman Bradburn has argued that positive and negative "well-being" might be separate dimensions (in McDowell and Newell, 1996, p. 191). A person might both suffer mentally and have well-being at the same time. This means that we should ask questions about both positive and negative states. However, if our theories of health, ill health, and well-being are well formulated and thought through, this should not be necessary. Health and ill health, well-being and suffering would be opposites.

However, psychological factors might exist that make people answer questions about well-being (for example: Have you lately felt up and in high spirits?) and suffering (for example: Have you lately felt downhearted and blue?) differently, and which might result in a slight bias.

For the sake of illustrating different possibilities, I have here, despite the possible problems with this approach, chosen to formulate the three kinds of questions both in a positive and in a negative mode.

11. Creating an Instrument: Example

I would now like to spell out in some detail what an instrument of this kind could look like. My model suggests some possible questions derived from the three categories outlined above.

A. Questions About General Health

Following the assumption that people have basic vital goals, we can ask questions about the ability to reach these goals. The stipulated basic goals are "to work," "to take care of one's home," and "to have some basic social contacts." Nordenfelt, Petersson, and Liss (1994) have suggested that a fourth area exists where people often have vital goals; I will call this category "leisure activities." Even though these are not part of my original list of basic vital goals, I will tentatively add them to this list, since most people can be expected to have some kind of leisure activity.

Another category of questions can be included here: questions about the general ability to do anything at all, like getting up in the morning or making decisions. Some might argue that decision-making is a specific mental ability, and thus belongs to the next category. For now I will include this question under general health. Possibly, questions like numbers 6–7 belong to the next category, not to this one. Is sleep, for example, a general ability or a specific mental one?

Let me now exemplify with some possible questions:

Have you lately

(1) been able to take care of your home?
(2) been able to manage your work (studies, or similar tasks)?
(3) had the energy to see your friends?
(4) been able to pursue your usual leisure activities?
(5) been able to make significant decisions?
(6) been having a hard time getting up in the morning?
(7) been having problems falling asleep at night?

B. Questions About Specific Mental Abilities

Here I try to give examples of different mental abilities: cognitive, conative, and emotive. Note that the questions about the emotions in this section have to do with being able to experience them, not with what emotions the individual has experienced during a past period of time. A connection between the two exists, since in order to know whether the individual has the ability to experience an emotion, he or she has to have had at least one such experience recently.

Have you lately

(1) been able to concentrate on what you are doing?
(2) been able to experience positive emotions?
(3) been able to think clearly?

(4) felt motivated to do the things you usually do?
(5) been having problems remembering things?
(6) been having less patience than usual?
(7) been having problems finding the energy to do things?

I assume here that patience is a specific mental ability.

C. Questions About Well-Being

These questions have to do with the experience of well-being or suffering during a past period of time. We can divide the questions into two basic categories: those having to do with moods and those dealing with emotions. Some questions might also have to do with reflections or judgments that are connected to the moods or emotions.

Have you lately

(1) felt optimistic and hopeful?
(2) felt good about yourself?
(3) felt on the whole that you were doing things well?
(4) been feeling downhearted and depressed?
(5) lost confidence in yourself?
(6) felt worried and anxious?
(7) been thinking that life is not worth living?

Note that the last question has to do with a judgment, not with a feeling. The third question might also be seen as being about a judgment, even though formulated as having to do with a feeling about life.

12. Conclusion

Many of these questions will be familiar, as some of them have been taken from well-known instruments. Why, then, the need for a revision of old instruments or the construction of new ones? The reason is that we have to be explicit about what it is that we want to measure. Only then can we formulate questions and be sure that they are related, conceptually or causally, to the goal we have for the treatment in question. Conceptual clarity increases validity. Given that the questions are logically related to the basic concept of what we want to measure, or have been shown to be causally related to the achievement of the goals in question, the validity is as reliable as the definition of the concept used.

I have claimed that health is the primary goal for health care, and I have chosen to use one general theory of health. I have also shown what kinds of questions can be asked given this theory. But the major point here is not that this is the only possible theory; different theories can be used. The point is that when we construct an instrument, we have to start with a theory of general health, which must be argued for, and that the questions used in the instrument have to be overtly related (conceptually or causally) to the theory chosen.

Part Three

HEALTH AND ILLNESS FROM A PHENOMENOLOGICAL POINT OF VIEW

Six

HEALTH AS RECEPTIVITY: A PHENOMENOLOGICAL INTERPRETATION OF ALLOSTASIS

Jennifer Bullington

In this chapter I present a concept from stress research, the so-called "allostatic principle" (or "allostasis"), and analyze this medical construct in light of a being-in-the-world perspective. I make use of French phenomenologist Maurice Merleau-Ponty's idea of structure and structure transformation to existentialize allostasis, to put it back into the world. I will thereby replace traditional medical conceptualizations of objective, physiological, third-person processes taking place in a mechanical, material body with a lived-body perspective. After presenting a possible Merleau-Pontian interpretation of allostasis, I conclude with a tentative suggestion that the phenomenological analysis of allostasis points towards a definition of health as receptivity.

1. The Concept of Allostasis

Neurobiologist Peter Sterling and epidemiologist Joseph Eyer introduced the term "allostasis" in 1981 (Sterling and Eyer 1981a, 1981b), as a new paradigm for understanding bodily patterns of arousal and adaption. Allostasis refers to the regulation of complex physiological processes in the body over time. We have known for a long time that the ability to adjust internal bodily processes in relation to what is happening around us is absolutely necessary for our survival. The autonomic nerve system, the hypothalamic-pituitary-adrenal axis, and the cardiovascular, metabolic and immune systems all participate in our constant adaptation to the environment. The allostatic principle is a modern improvement upon Walter Cannon's (1935) principle of homeostasis from the 1930s. This principle described the way that the body appears to strive toward stability, toward a steady state. Homeostasis included a narrow set of variables—blood pressure, hormone levels, and so on—that must be maintained in order to keep the organism out of peril.

The criticism of homeostasis was that these so-called "normal" values or parameters never operate in real life. The idea that a set of normal physiological values would exist toward which the body necessarily strives proved to be a fiction. Researchers learned that they needed to know the context of a situation to understand the normal or abnormal value of a given variable. For example, high blood pressure in some situations is quite normal, and can be maintained in an appropriate situation for a long period of time without any damage to the system.

While homeostatic regulation is achieved through the use of local, negative feedback loops, allostatic regulation involves the entire body, orchestrated centrally by the brain. The brain can and does override local homeostatic feedback mechanisms that would oppose its commands. Allostatic functioning can anticipate needs and prepare the body for action before a real situation arises. For example, even before I get up from my chair, my body is prepared for this event in terms of its blood pressure, muscular tone and so on. Homeostasis can correct problems that have already arisen, but the most advantageous time to adjust for needs is before they arise.

To recap: Advocates of homeostasis claimed that the body's internal milieu is constant (within normal parameters) as a result of local, self-correcting, negative feedback mechanisms of the organ or system involved. The allostatic principle, however, denies the existence of a "normal" value or state that does not relate to a behavioral context. High blood pressure over a long period of time is both relevant and adaptive if one is engaged in a stimulating and challenging situation. The key to understanding an organism's well-being is described in terms of allostatic flexibility and adaptiveness to concrete situations instead of the homeostatic striving toward a fixed set of abstract, "normal" values.

This total adaptation of the organism to the situation (both simultaneously and in an anticipatory mode) is achieved through the holistic brain-body system, where the brain can override local negative feedback loops for the good of the whole organism. The allostatic model brings the meaning and signification of events into the arousal equation, maintaining that the body's physiological responses to the challenges of life are mediated through the cognitions and interpretations that we make. Allostasis is thus an improvement upon homeostasis. Allostasis takes into account the situation-bound context of arousal and adaptation and the importance of the cognition, interpretation, and evaluation of the situation. What is normal and abnormal in the body is a matter of interpretation.

But if we dispense with normal values for health and well-being, this does not mean that the organism can tolerate any level of arousal for any length of time. Bruce McEwen's conception of "allostatic load" describes the relationship between neuroendocrine activity (levels of responsiveness and arousal) and health risks (McEwen and Stellar, 1993). The allostatic load is the cumulative strain on the body produced by the repeated ups and downs of physiological response. This is a typical stress theory idea: that repeated and prolonged physiological activity eventually takes its toll on an organism. The metaphor of balance often functions as a model for the healthy, adaptive organism. To maintain balance over time is to be able to turn on a response, when necessary, but also to turn it off when the situation has passed. Several types of allostatic load have been described in published research; the most common forms are presented below.

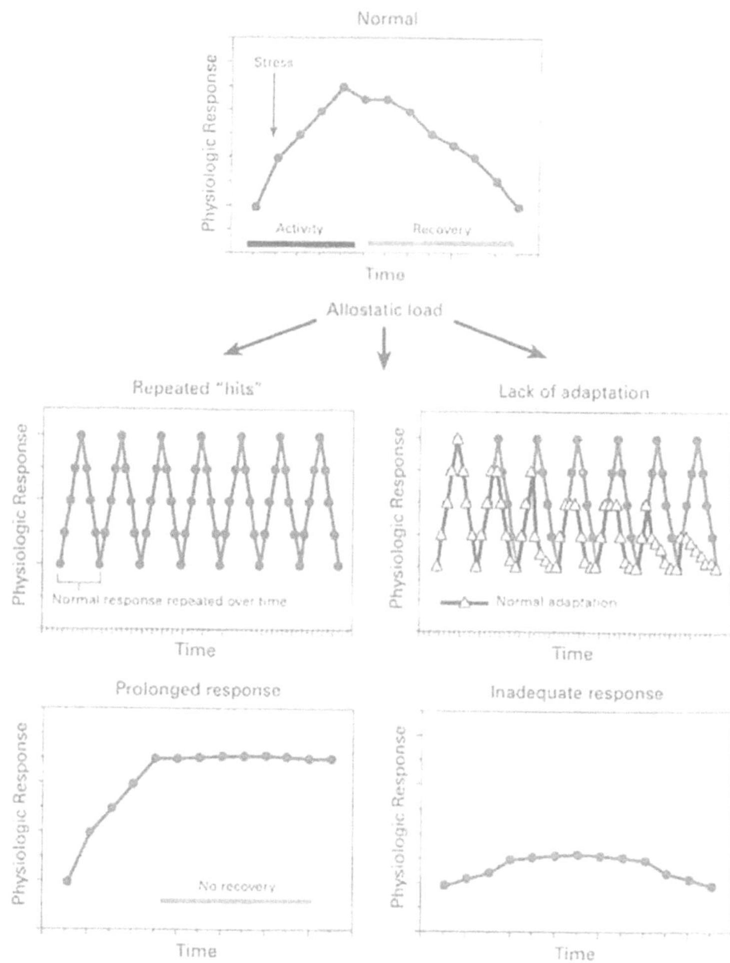

Figure 1. The New England Journal of Medicine, 338, p. 174.

The top figure illustrates a normal allostatic response, showing a stress reaction initiated by a stressor that is sustained for an appropriate time and then turned off. The figure at bottom right represents the most common form of allostatic load, so-called "repeated hits." This is what lay people usually refer to by the term "stress." In this case, the person is again and again subjected to situations where he or she must "turn on" a stress response.

Another form of allostatic load, called "lack of adaption," involves sustained arousal in relation to repeated stressors of the same type, resulting in prolonged exposure to stress hormones. Where most people would eventually adapt to such a situation without turning on the stress response, subjects experiencing this disorder cannot adapt. They continually become alarmed at the situation, despite its familiarity. "Prolonged response" is the inability to shut off allostatic responses after a stressful situation is terminated. One behaves as if the stressful situation were still going on.

Finally, "inadequate response," depicted in the last table, illustrates how the failure to respond adequately in one system results in compensatory overactivation of another system. For example, insufficient cortisol secretion leads to an increase in inflammatory cellular activity. All of these reactions may be said to be inadequate, in different ways. They are non-adaptive in the sense that they do not result in a satisfactory mastering of the situation. To master a stressful situation, according to this paradigm, means being able to rise to the challenge (turn on the response) when necessary; to adapt to the challenge if it is repeated—that is, to normalize the situation so that it is no longer experienced as a challenge or threat; and, finally, to be able to turn off the response when it is no longer needed.

Some stress research has shown that in cases of prolonged allostatic load, the body may become addicted to its own stress hormones (Friedhoff and Miller, 1983; Reisine and others, 1986; Yamamoto, 1985). When the body's arousal level has been high for a long period of time, relaxation may be an unpleasant experience for the subject. The higher the chronic levels of stress hormones, the more a person's mood may be dependent upon keeping those levels high. Thus, chronic arousal has a tendency to become self-maintaining. Researchers have proposed that this mechanism is the physiological basis for Type A behavior such as workaholism.

Much of the work in so-called "appraisal" stress theory, which started in the late 1960s with the work of John Mason and others, points to a link between psychosocial factors and physiological processes. However, this relationship is not as clear-cut as one might imagine. Positive social contacts are said to reduce neuroendocrine activity. The more diverse someone's social network is, the greater that person's resistance to disease is. When social relationships are disrupted, we often see a rise in age-specific morbidity. But empirical studies show contradictory findings. For example, one study of people anticipating a public speaking engagement (Kirschbaum and others, 1995) showed that the presence of a socially supportive companion was only beneficial for men; for women, the presence of a partner was associated with a stress response.

Earlier studies have indicated that significant associations between emotional support and lower endocrine activity are strongest among men (House and others, 1982; Kaplan and others, 1988). We can conclude from those studies that women benefit less from social ties because such connections "cost" more for women, who tend to be more emotionally open and responsive, vulnerable to negative social interactions, and so on. Also paradoxical is the finding that while many community-based population studies have shown that the higher the subject's social status is, the lower his or her stress variables (cortisol, heart rate, blood pressure, etc.) will be, other research indicates an association between high social status and high levels of morning cortisol—that is, a high level of morning stress response (Bourne and others, 1968; Brandtstadter and others, 1991). This is intriguing, since high stress levels are usually associated with lower social status, and conversely, high social status correlates to low stress variables.

What could this finding mean? Brandstadter and colleagues interpret this in terms of what they call "adaptive coping"—a positive readiness to meet and cope with novel and challenging situations. This would mean that we can associate coping with high cortisol levels, which have been one of the traditional measures for negative stress and inability to cope. Accordingly, high levels of stress hormones in the morning need not necessarily be associated with the experience of stress or negative health outcomes. Such findings point to the need for further investigation. But I do not believe that empirical correlation studies will help us properly understand stress as an embodiment phenomenon. The phenomenon of stress is about the mind-body-world system. To achieve a deeper understanding of stress, including some of its paradoxes, I believe that we need to re-interpret the stress theory paradigm in a new way. I will now present an alternative way of thinking about stress.

2. Merleau-Ponty's Notion of Structure

Maurice Merleau-Ponty (1942/1963; 1945/1962) is the phenomenologist who has brought the body back into philosophy. But the "body" he refers to is the lived body, not the body understood as an object or described in terms of natural scientific constructs. Merleau-Ponty was critical of what he called Husserl's remanent idealism. He was equally concerned about the influence of natural scientific thinking on how we perceive ourselves and our relationship to the world. We have a cultural heritage of dualistic thinking, of splitting up the world into an objective material domain and a transcendental, subjective realm. We furthermore tend to interpret human experience in natural scientific terms. The unhappy result of this heritage is that we have learned to consider our bodies as "objective" material bodies or a collection of physiological and biomechanical processes. We are likewise convinced that the world is "really out there" exactly as natural science has described it, and for that reason we tend to interpret our experience of the world in objectivistic terms. For example, we take for granted that our perceptual experiences are the result of external stimuli impinging upon our sense organs.

The phenomenological tradition has always pointed out that however interesting and valuable these natural scientific theories and findings may be, they are constructions and in no way causally responsible for, nor temporally prior to, experience. They are derived from life, our world experience, and therefore cannot be placed prior to experience. Following this critique, Merleau-Ponty's ambition was to distinguish philosophical thinking from both idealism, with its dualistic implications, and what he called "empiricism," or natural scientific thinking. He felt that these ways of thinking exerted too strong an influence upon philosophy and psychology.

Merleau-Ponty gave some credit to Gestalt psychology for coming up with the notion of form, but he criticized the philosophical realism underlying Gestalt psychology. He found this unsatisfactory, and felt compelled to further develop the Gestalt discovery of form himself, which he did in his *Structure of Behavior* in 1942. His notion of form—which he called "structure"—is neither "out there" (in the world, empirical) nor "in here" (entirely constituted by the subject, idealistic). Structure for Merleau-Ponty is an ambiguous dimension that is neither thought nor thing, but the meeting between the two. Structures are "in-between" in that they can only exist for a subject situated in a concrete life world.

Can we define structure more precisely? Structures are the bodily, psychological, socio-cultural patterns that lead us in our experience of the world. Without these structures, we would not "have" a world. Although they are necessary for us to grasp and understand our world, they are themselves constantly formed by our experiences. Our contact with the world modify these structures. Yet they are at the same time that by which we can recognize, organize and understand the world, on all levels.

According to Merleau-Ponty, human beings are constantly engaged in an ongoing dialogue with the world. This is a kind of question-and-answer dialectic, which both develops and constrains our understanding and experience. Structures allow us to feel at home in diverse situations, to recognize and implement habitual ways of perceiving and thinking. We thereby free us to pursue the possible, the new and the unexpected. We do not need to relearn our mother tongue every time we use it, nor do we have to pay undivided attention to our bodies when we perform habitual movements. However, structures are not only stable and sedimented, they are also spontaneous and capable of transformation, since no structure will always and in every way fit every possible situation.

A final essential characteristic of structures is that they seek harmony. A successful structure is one that resolves the tension that inevitably arises between man and the world. When confronted with an intellectual problem, for example, one searches for a structure (in this case, a higher-order, cognitive one) that solves the enigma. When surprised by something unexpected or unexplained, we immediately try to fit this new and different element into something already known. Or, if we cannot succeed in that, we are cast into the tumult of "making sense" out of it. The process of making sense out of things is the transformation of habitual structures in order to allow this new thing or event to become a part of the intelligible world.

The dialogue between man and world goes something like this: Something presents itself as either familiar or unfamiliar. If that thing is familiar, the person will quickly respond to it with habitual, sedimented structures. If it is unfamiliar, a tension arises, a tension that can be experienced at several levels: bodily, psychic, social. This tension calls for a structure transformation in order for harmony to prevail between the given and the intending. The world thus presents us with a rudimentary organization, unfinished and beckoning, which calls for our participation.

The world is often presented to us in an ambiguous fashion. Events and things are not fixed and finished in the world. Things and situations become, they unfold, they raise questions and they disturb. Consider these scenarios: (1) Standing on a dark, empty street, I wonder, "Is it a man over there in the shadows, or is it a garbage can? Let me get a closer look." (2) Walking up an incline, one moves the body slightly forward. (3) You accidentally find out that your partner is having an affair, giving rise to a variety of serious questions, such as, "Who is he, and what are we, now?"

All these changes are structure transformations. The transformation of structures is the ongoing process of being alive in a human world. We are continually challenged by the world to further articulate, deepen and modify our understanding. The more we deepen and articulate our understanding, the more of the world is shown to us. We are already immersed in the world, directed toward it. We are continually asking questions of it, being questioned by it, sedimenting the world in layers of ready-made significations. But we are also prepared to overturn comfortable, habitual structures and transform the signification of the given when necessary.

3. A Merleau-Pontian Interpretation of Allostasis

Now, what can a Merleau-Pontian interpretation of allostasis bring to light? I would claim that allostasis already describes a being-in-the-world phenomenon, since its understanding of stress and adaption takes into consideration the meaning and context of behavior. But the allostatic model lacks a being-in-the-world framework for understanding the significance of what it describes. Many of Merleau-Ponty's ideas about structure and structure transformation can be applied to the allostatic model of arousal and adaption.

One finding related to the anticipatory function of allostasis is that predictability and sense of control result in low arousal levels—that is, low stress. When the subject automatically knows what is coming and what is expected, the system is not aroused. In Merleau-Pontian terms. We can understand this as an example of sedimented structures as harmonious patterns of intelligibility. The structure fits the given; transformation is not called for at any moment. The smooth, automatic nature of this kind of experience is a part of everyday life. On a bodily level, we are hardly ever challenged to seriously modify a sedimented structure. For adults, this only happens when we have to learn a new motor task. One of the reasons a funhouse makes us laugh could be that we find it peculiar that we cannot use our sedimented body structures in an environment where we expect that they should work. What we anticipate bodily and what is actually called for is out of sync.

In a similar fashion, we use sedimented structures on a psychological level until something arises that calls for a transformation. This may occur when hidden or not fully disclosed sides of a situation suddenly break through, demanding on our part renewed attention and sense-making procedures. Exploiting this moment is often a part of the suspense of the thriller film, when the audience sees the imminent moment when the situation will be dramatically overturned before the characters in the film see it. When that moment finally does come, the entire situation "explodes." Then everyone has to transform their structures in order to grasp and accommodate the new scene: The best friend was the killer.

The finding that the brain-body system can become addicted to chronic arousal can be understood in Merleau-Pontian terms as a preferred pattern of disharmony. The subject has for one reason or another "chosen" to maintain the experience of the world as arduous, difficult and out of control. The maintenance of chronic arousal becomes comprehensible when it is seen as an expression of one way of being-in-the-world. The person prefers to live as tension, battle and resistance.

Instead of wondering about the (physiological) mechanisms that maintain chronic arousal, the phenomenological point of view would see this addiction as a possible world. The chronically aroused person either actively seeks challenges again and again ("repeated hits") or does not transform structures in order to accommodate the new and challenging ("lack of adaption"). In both cases, experience does not sediment and become part of a smooth flow of habituation and intelligibility. The chronically aroused person is living and expressing a style of existence.

Instead of examining the objective body by noting heightened hormonal levels or other indicators, the phenomenological perspective regards the mind-body-world situation as an integral whole. Seen in this way, allostatic load is characterized as the inability to transform structures at the appropriate time or, in other cases, the inability to sediment structures when necessary. Thus, the entire person-world encounter does not "fit"; the situation is not just a matter of inappropriate physiological activity.

Each different form of allostatic load as defined by McEwen could easily be interpreted as a being-in-the world phenomenon. "Repeated hits" could be understood as a refusal to live life primarily through sedimented structures. The person seeks out and creates disharmonious experience. The preferred experience is the tumultuous process of trying-to-make-sense-of-X. To sediment structures and to live through them is to get beyond this tension and "relax" into habituation. Repeated hits keep the individual in a constant state of transformation, a continual being-on-the-way-toward-mastery, a striving that does not come to rest.

The case of "prolonged response" can be understood in terms of a disturbance in the temporal flow of experience. Prolonged response is a way of holding onto an experience as if it were still the case. The problem here concerns moving on. The person maintains a structure (a mind-body-world position) as if it were still applicable, despite its inadequacy to harmonize with the present situation. For example: I am still fighting with my boss, even though I am in reality at home trying to get to sleep in my bed. Here, the thrill of the sense-making experience is not central, but rather the inertia of staying. Prolonged response could be seen as a way of keeping the future from arriving. To understand the phenomenon of prolonged response, we need to ask what is being prolonged, and why. The answer to these questions cannot lie in the analysis of hormones and brain processes. They exist in the being-in-the-world of the person experiencing this phenomenon.

4. Health as Receptivity

What can a being-in-the-world interpretation of allostasis tell us about health? The allostatic model speaks of health as responsiveness. Therapeutic advice tells us to reduce arousal. I would prefer to speak of health as receptivity. The term "responsiveness" brings to mind the old stimuli-response associations, the idea that stimuli impinge on the organism and must be responded to. We respond to the world, but we also actively constitute the meaning and signification of the world, on all levels at all times. If we replace the stimuli-response model with being-in-the-world, we may characterize health as the ability to be open to the world. In order to receive or take in the world—the lived situation—the subject must be able to sediment structures into a pattern of coherence and signification and to transform these structures and break sedimentation when necessary.

Health as receptivity means that we are healthy when we are open. Being open does not mean just passively "taking in" (which would only be another way of describing allostatic "responsiveness"), but also involves an active, subjective questioning of the world, with a spontaneous interest and appetite for new aspects and experiences. This double-edged openness or receptivity, which entails an accepting and taking in as well as interrogation and discovery, is a prerequisite for all learning and development. If we were not open, we would never see anything new, think a new thought or discover anything exciting and different. Silvan S. Tomkins (1982) lists as one of our main affects the desire to experience novelty and resist boredom. Such desire is surely a basic characteristic of human beings.

Returning to the notion of stress, we know that a person can experience stress when s/he has too much to do, and too little to do. Extremes are said to be detrimental to health. The notion of balance is one of our major metaphors for health. The ancient Greeks considered balance among bodily fluids or between different universal elements to be essential for maintaining health. The Chinese model of health and illness, with roots as far back as 2800 BC, rests on the notion of two complementary cosmic powers, *yin* and *yang*, whose harmony is essential for the order of the cosmos and for the health of the individual.

Modern theories of health point out the importance of balance or equilibrium among an individual's ambitions, environment, competence and goals (Nordenfelt, 1987; Pörn, 1993). Since balance appears to be such a central factor in our intuitive understanding of health, the next question to address is: Can we understand health-as-receptivity in terms of balance? I will now consider balance in two dimensions: balance in the current, "now" situation, and balance over time.

I will begin with balance in the current situation. I have characterized health as being open. This openness entails a "taking in" of the given and being able and willing to seek out and discover new and different aspects—aspects which would then require (major or minor) structure transformation to be assimilated. To be in balance in everyday, routine situations (riding the subway, grocery shopping, grading exams, etc.) is to allow sedimentation to be the dominating mode. In everyday situations, sedimentation will be appropriate. The intending fits the given. I will not normally be aroused or vigilant. If I find that I must concentrate on myself or on my body in such routine situations, I experience myself as unwell. Examples of this would be a phobic trying to shop in an enclosed shopping mall or a newly operated person attempting to move his or her painful body through familiar movements.

But even in the most common sedimented situations, the healthy person will never be completely immersed in the habitual. Being open is being willing to be surprised. In fact, one of the major benefits of sedimentation is that it gives us energy and mental space so that we can always enrich our experiences. The ease of routine frees us for the new and different. So balance in the now of everyday situations means that although sedimentation is the dominant mode, we exhibit a readiness to break routine and experience the new and different. Being completely closed in sedimentation in an unhealthy way can be exemplified by neurosis. Severely neurotic people cling rigidly to habits and routines ("the way it is") and find it extremely threatening to allow any change or alteration in the pattern of intelligibility that they have constructed for their survival.

I will now consider balance in terms of the now of uncommon, challenging situations. The boss asks me to come into her office. This situation is not unusual, as she often calls me in from time to time in order to discuss the projects for which I am responsible. I sit down, expecting our usual type of meeting. However, I immediately sense that something is different. Her face looks different, the atmosphere feels strange, and my "conversation-with-the-boss" sedimentation begins to move into a readiness for something new. She tells me flat out that the company is going to have to let me go. My work is not up to par, and they have decided to give my projects to someone else in the firm. In such a situation, the dominant mode is structure transformation: I must try to make sense out of this. A million questions swirl about in my mind. What did I do wrong? Has some person in the office out-maneuvered me? Who? What am I going to do now?

These questions, and the answers to them, are the work of structure transformations. Striving to make the incomprehensible comprehensible corresponds to what we do on a bodily level when we have to move our bodies in an unfamiliar way: I must get there, but I don't quite know how. But just as adult bodies have a repertoire of movements to draw upon, so we as psychological and social creatures have at our disposal a background of psychological "movements" that we may try out. What, then, constitutes balance in the case of the uncommon, challenging situation?

First, we must be able to take in the new unexpected situation. In the example above, to keep on applying my usual "meeting-with-the-boss" structure would be inappropriate. I must be sensitive and attuned to the challenge that faces me. An unhealthy, unbalanced person would not even hear or understand the message that the boss delivers. After the situation is "taken in," the next step concerns performing the work of transformation. This means that I am able to use what I have accumulated in experience to try to make sense out of the challenge. For example, to sit and stare at my boss with my head completely empty or full of nonsense would be a blockage of transformation. The work of "making sense" requires that I can understand what is happening.

To understand, or make intelligible, is to get a structure which will accommodate the given. My new structure may turn out to be something like: (1) I have not been performing well at work, although I was unaware of this fact because of a variety of ongoing personal problems; (2) I have been blind to intrigues that have been going on around me in the office; (3) The competition in modern working life is much harder than I had realized. The challenging situation balances the dominance of the transformation mode with maintaining strong enough connections to sedimentation that I may begin to construct a new intelligibility based upon the old. Were all ties to usual sedimented patterns of understanding blocked, we would find ourselves in a state of chaos, psychosis, or somatization instead of transformation.

Health as receptivity can also be understood in terms of balance over time. Whereas McEwen's "allostatic load"—and much of stress theory in general—refers to imbalance as an accumulative strain on the machine body, the wear-and-tear on a physical thing, I would prefer to speak of balance and imbalance over time in terms of the mind-body-world system. Balance over time means a shifting over time between habituation and challenge, although the exact proportions of this shifting will vary from person to person.

The salient point is that an ongoing flow exists from positions of more or less stable sedimentation to excitement and challenge. If we return to the neurotic patient, we see that much of that person's suffering has to do with the way in which his or her rigid style is constantly under attack, both from within and without. People can hardly master and live complicated human life with only a handful of stereotyped, rigid structures available. And even if the neurotic does manage to restrict his or her living space to a small circle of predictability and routine, the monotony and lack of vitality and spontaneity makes this life difficult to live.

Longings, desires and all other forms of human striving hammer away from within, while the world beckons and challenges from without. On the other extreme we would find the thrill-seeker or risk-taker, the gambler, the adventurer, the mountain climber, the Don Juan. This personality style is all challenge, with little tolerance for routine and habits. As soon as life becomes predictably boring, the person is off again looking for stimulation. This imbalance over time is the inability to rest. Just as the body has its rhythm of activity, rest and sleep, so we can imagine the rhythm of the mind-body-world system, consisting of a pendulum movement from habituation and sedimentation to challenge, excitement, and structure transformation. Obviously, different people will have different preferences for just how much of their life they wish to be habituating to contra challenge. The relevant point is that balance over time means that we move back and forth on this continuum.

5. Concluding Remarks

To be in good health, then, is to be able and willing to sediment and transform structures. Traditional Chinese medicine has a similar conceptualization of health, as living life open to the flow of *Qi* (life energy), which moves continually inside and outside of us. Ill health and disease are attributed to a "blocking" of the flow of *Qi*. The implication of health as receptivity for Westerners is perhaps to focus upon how we as human beings depend upon a balance between sedimented structures (allowing for stability) and the capacity to transform structures when necessary (allowing for change and growth).

Seven

A PHENOMENOLOGICAL ANALYSIS OF THE CONCEPTS OF HANDICAP AND ILLNESS

Fredrik Svenaeus

1. Introduction

This paper is an attempt to draw distinctions between the phenomena of handicap and illness from a phenomenological point of view. A proper distinction is necessary because the handicapped most often do not consider themselves to be unhealthy despite their functional defects. A blind man talks about falling ill when he catches a cold, but he does not consider his everyday existence as unhealthy just because he is not able to use his eyes.

What use can be made of such a distinction between handicap and illness, and why would philosophy—in particular phenomenology—be the proper tool for such an analysis? We must address these three questions briefly at the outset. Conceptual analysis has formed a core part of philosophy ever since Socrates strolled the streets of Athens and put nagging questions to his fellow citizens about the nature of different phenomena. In the Platonic dialogues, these questions often extend into the area of medicine (Petersson, this volume).

The need for conceptual clarity in medicine and health care is just as strong in our time as it was in those days. Such distinctions are useful for doctors and other health-care personnel, just as for the politicians developing our health-care policy. They are also relevant to the people afflicted by different ailments, not least to the handicapped, who have often had to fight for the right to enjoy equal opportunities in society. Thus, we should not view good philosophical analysis as an ivory-tower project, but as a kind of reflection that dwells in the same region as the phenomena it investigates.

The question of why the philosophical school of phenomenology is so appropriate for studying the philosophy of medicine is harder to answer briefly (Svenaeus, 2001). The common root of all phenomenological analysis comes down to Edmund Husserl's claim, in the beginning of this century, of the necessity of going back to "the things themselves" and the way they are presented to us in order to find a solid ground for philosophy and science

(Spiegelberg, 1982). We can find the conceptualization of appearance—the *eidos*, or meaning, of experience—according to Husserl, in the phenomena themselves by adopting a certain mode of philosophical attention. This project has been taken up and developed in many different ways, some of which I will address in this paper. The basic concept of phenomenology is precisely explication of the meaning of everyday experience.

Illness and handicap form essential parts of this everyday world and are thus open for and suitable to phenomenological analysis. We can view the borderline between normal and abnormal experience as part of the founding ground of phenomenology. Husserl's French successor, Maurice Merleau-Ponty for instance, used the experiences of people suffering from different neurological disorders to get hold of the structure of normal experience (1962).

2. Philosophy of Health

Health theory is not a new subject in philosophy. During the last three decades we have, however, witnessed what might be called a revival of the philosophy of medicine and health. For different reasons, the insight that the success of medical science by itself will never provide us with answers to all questions in health care has become more widespread. The exploding interest in medical ethics has been the most obvious sign of this need for philosophy in the field of medicine. Medical science itself does not provide us with a language in which to carry out a philosophy of medicine and health. Medicine needs philosophical theory and vocabulary to meet these demands. Consequently, several philosophers have recently approached health theory and offered different suggestions for how to draw a conceptual map of health and other related phenomena.

Most of these theories of health tend to meet with problems when it comes to pinpointing the borderline between handicap and illness, since the criterion chosen to demarcate healthy individuals from ill ones also tends to exclude the handicapped. The handicapped are automatically defined as ill by the theory. In order to see this more clearly, let us review two influential health theories—those of Christopher Boorse and Lennart Nordenfelt.

Boorse's theory defines ill health as the presence of a pathology that causes biological malfunction in its bearer (Boorse, 1997). Normal biological functioning is defined statistically on a species level. The biological organisms of handicapped persons always present us with some dysfunctions, and the handicapped are therefore defined as unhealthy. The theory does present a way of defining the difference between handicap and other types of ill health, however, by pointing to the different forms of pathologies or maladies that the ill, versus the handicapped. are suffering from. The handicapped are not diseased; they do not suffer from processes that interfere with normal biological functioning. They are, instead, suffering from congenital defects, impairments, and injuries. The maladies that they suffer from are not biological processes, but biological states. Still, because these states interfere with normal functioning, the handicapped are looked upon as unhealthy in Boorse's biostatistical theory.

In recent years, persons suffering from maladies other than diseases have often preferred the term "disabled" over "handicapped." This change in terminology appears to point toward a theory of health and illness that is focused on the subject as an individual who inhabits the world through his or her actions rather than on biological functions. Nordenfelt builds his holistic alternative to Boorse's theory on the notions of ability, action, goal, and circumstances, instead of on biological-statistical normality. The healthy individual is able to realize his or her vital goals given standard circumstances. We will not here dwell on all the difficulties that specifying vital goals and standard circumstances presents us with; Nordenfelt has taken care of most of the objections that his theory might meet with in his studies, notably in *On the Nature of Health* (1995a).

The crucial point is that Nordenfelt's definition represents a descriptive and normative attempt to formulate a theory of health. In contrast to Boorse's theory, in which health is a theoretical, scientific concept, health in Nordenfelt's theory depends upon the evaluations made by human beings who consider what is vital and standard to them. This does not make the theory "subjective" in the sense that every individual could choose which standards are equivalent with a healthy situation for him or her, but it stresses that health is partly dependent upon assessments made by human beings interpreting their own situation.

Instead of discussing Nordenfelt's theory in detail, which I have done elsewhere (2001), I will focus on how that theory, just as Boorse's, defines the handicapped as ill rather than as healthy. This must be so, since such people are disabled—that is, unable to realize their vital goals given standard circumstances. The handicapped demand special circumstances and extraordinary equipment, such as wheelchairs, blind sticks, or hearing aids, to carry out their life projects. Thus, the focus on ability in Nordenfelt's theory—despite taking into consideration the whole situation of handicapped people and pointing out the ways in which they are disabled—automatically defines the handicapped as ill.

The only way to differentiate between the ill and the handicapped in Nordenfelt's theory would be to focus on the different forms of biological maladies the individuals are suffering from—that is, either a process (disease), or a state (defect, impairment, or injury). This is similar to Boorse's effort to conceptualize the difference between handicap and disease. But Nordenfelt might offer another answer in response to my criticism. He could point out that everybody is handicapped in respect of some external circumstances—I was handicapped when it came to the activity of sunbathing this summer in Sweden, for instance—and that handicap always regards the disability in respect of a set of circumstances in which to carry out the activity in question.

This argument, however, operates with a different concept of handicap than the one I have in mind. The term "handicapped" as it is used here covers only the group of people that we today view as medically handicapped in our society. They are in most cases identified by way of a biological malady. It might be that Nordenfelt wants to give up this concept of handicap altogether and instead focus upon disability. This decision would find strong support among many of the handicapped and their organizations. I find this move away from the notion of biological handicap tenable, but I envisage a different path than Nordenfelt's. The problem, as I see it, is that being disabled in respect to vital goals under given standard circumstances still equals unhealthy, in Nordenfelt's theory.

Before I turn to my proposal for a possible distinction between illness and handicap, I would like to highlight that my criticism of the health theories of Boorse and Nordenfelt on this point is not that the handicapped should in every case be defined as healthy. Obviously, some handicapped suffer from maladies so severe that it would be contraintuitive to regard them as healthy—think of tetraplegics, for instance. My criticism is, rather, that both theories automatically define the handicapped as ill without offering the possibility of distinguishing between handicaps that can be compensated for in such a way that the bearer can lead a healthy life, and those that cannot. The key issue for many types of handicaps is that while the bearers do not live lives that are normal in the sense of being like everybody else's, their lives are healthy in that they have managed to change their abilities, projects, and external circumstances in a way that makes their existence homelike.

To pinpoint the difference between illness and handicap, we need a theory that pays close attention not only to ability and action, but also to thoughts and feelings. One of my suggestions in this paper is that phenomenology provides us with such a language in its focus on lived experience—that is, the feelings, thoughts, and actions of the individual person living in the world. Such a theory of health will, just as Nordenfelt's holistic alternative, be normative, since it will fundamentally rely on the individual's interpretation and evaluation of his or her situation, and not only on a biological investigation of his or her body. But the physiology of the body afflicts and sets limits on the ways we are able to experience and interpret our situation.

To develop a phenomenological theory of health is therefore not meant as an attempt to replace biomedical research; in light of the successful history of modern medicine, that would be an absurd project. Phenomenology is meant to enrich our understanding of health in adding to the disease-level analysis a level that addresses the question of how physiological states are lived as meaningful in an environment.

3. Heidegger's Phenomenology

Drawing on my previous attempts to develop a phenomenological theory of health and illness in *The Hermeneutics of Medicine and the Phenomenology of Health* (2001), I will now try to find a distinction between handicap and illness working in the framework of Martin Heidegger's existential "being-in-the-world." Is the being-in-the-world of the handicapped different in any significant sense from that of the chronically ill, for instance, with whom they appear to share many predicaments? The governing idea of this analysis will be that the being-in-the-world of the handicapped generally can be understood as a homelike, although non-standard, being-in-the-world.

Heidegger, in his first main work, *Sein und Zeit* from 1927 (1986), widened the domains of phenomenology. Whereas Husserl was mainly concerned with epistemology—the theory of knowledge, with an emphasis on the theory of science—Heidegger rather focused on the everyday world of being and understanding. His phenomenology is accordingly what he calls a "fundamental ontology" that investigates different modes of what it means to be, rather than what it means to know.

As his starting point, Heidegger takes, instead of the subject of knowledge, what he calls *Dasein*, the "being-there" of human existence. This being-there means that we are situated or "thrown" into the world that we live in. We are always already there (*da*), involved in daily activities. But the term *Dasein* also signifies that we have a relation to our own existence in asking what it means to be there at all (rather than not to exist). *Dasein* is the only being that understands its predicament and asks the fundamental ontological question of what it means to be. Therefore, philosophy has to start with an analysis of the understanding human being—*Dasein*. This is the only way to start, since every other being-in-the-world attains its meaning through the understanding of *Dasein*. Human understanding as a "being-there" in the world is, accordingly, the starting point for philosophy as a phenomenology of "everydayness"—an investigation of the everyday forms of understanding carried out by human beings.

According to Heidegger, when we study our relationship to the world around us, we should not view the world as a collection of objects outside of consciousness toward which we are directed by way of consciousness (this was Husserl's basic model). We should instead study the "worldliness" of the world: the way we are in the world, giving it meaning through our actions—the world being nothing other than a cultural, intersubjective meaning-structure lived in by us and, ultimately, a mode of ourselves. Human understanding is, for Heidegger, consequently, always a being-there in the sense of being-in-the-world. The hyphens indicate that *Dasein* and world are thought of as a unity, not as subject and object. The world is not something external, but is constitutive for the being of *Dasein*.

The concept "worldliness" in *Sein und Zeit* indicates that the structure of the world is built by the understanding actions, thoughts, and feelings of human beings situated in the world, not by any properties belonging to the world itself as a collection of objects (things, molecules, atoms). Heidegger can therefore write that worldliness is an existential—that is, something belonging to *Dasein*, to understanding human beings, and not to the world in itself.

The meaning-structures of the world are made up of relations not between things, but between tools. That is, the meanings of phenomena, according to Heidegger, are not primarily dependent upon how things look, but upon how we use them. This makes the connection between the structure of the world and *Dasein* more lucid. How could the world itself as something independent of human beings lead us to an understanding of the function of any tool? A tool always refers to its user. We will only learn what a hammer is by using it, never by staring at it (Heidegger, 1986, p. 69).

The relationships between the different tools are explicated as an "in order to" (Heidegger, 1986, p. 68). The tools in this way relate to each other; their meanings are determined by their places within the totality of relevance. You use a hammer to nail the palings, to raise the walls, to build the house, to find shelter from the rain, etc. You need not pay attention to these different levels of sub-goals at all times. Some of the sub-goals are never explicitly attended to, but are only revealed in a theoretical analysis of the activity.

The final meaning of every tool is the existence of a human being—*Dasein*. *Dasein* is the activity in relation to which every phenomenon (tool) takes on meaning. Heidegger conceptualizes the being-in-the-world, the "worldliness" of human existence, by stressing several aspects of this existence. Since these aspects belong to the only being that truly exists—*Dasein*—and not to things, they are called "existentials." Human beings exist: that is, they have a relation to their own being, and they are open to the world as a possibility for themselves. This openness to the totality of tools is a pattern not only of action, but also of thinking, feeling, and talking. But these three modes of being must not be conceived of as attributes of a subject, as qualities of a thing, but as a meaning-pattern that binds human beings and the being of the world together. They must likewise articulate a being that is not only contemplative but acts in the world, as the tool pattern makes obvious.

The three main existentials that Heidegger chooses in *Sein und Zeit* to make sense of our being-in-the-world are understanding, attunement, and discourse. These three existentials are thought of as intertwined: they always work together, inseparably, as an attuned, articulated understanding situated in the world. The choice of understanding instead of cognition makes a relation to action possible. Understanding can include the active and incarnated sides of life, which are essential to the analysis of health.

4. Developing a Phenomenology of Illness and Handicap

In my abovementioned previous study, I tried to show how healthy versus ill life can be explicated as homelike versus unhomelike being-in-the-world (2001). Homelikeness and unhomelikeness refer to two opposite modes of the attuned, incarnated understanding of human being-in-the-world. To be ill means to be not at home in one's being-in-the-world, to find oneself in a pattern of disorientation, resistance, helplessness, and despair, instead of in the homelike simplicity of normal life—think of having a bad cold or severe headache, for instance, or suffering from incipient diabetes.

The attunement of our being-in-the-world is a vital phenomenon to focus upon when we try to get hold of the difference between healthy and ill ways of being-in-the-world. To be healthy or ill is, however, not identical with just having a good or a bad feeling. Attunement is not the quality of an isolated human subject, but of a being delivered to the world; it is finding oneself in the meaning-structure of the world as an understanding existence. Hans-Georg Gadamer, one of Heidegger's followers in the phenomenological tradition, has, in a recently published collection of papers, characterized health in the following way:

> Health is not a condition that one introspectively feels in oneself. Rather it is a condition of being there, of being in the world, of being together with other people, of being taken in by an active and rewarding engagement with the things that matter in life.... It is the rhythm of life, a permanent process in which equilibrium re-establishes itself (Gadamer, 1993, pp. 144–145).

We recognize Gadamer's terms as derived from the phenomenology of *Sein und Zeit*, and we are able to understand this healthy, rhythmic being-in-the-world as a form of attunement that is not just a state that we feel within ourselves, but which nevertheless has a "tune," a feeling component to it.

The phenomenon of illness in many ways appears to be easier to get hold of than the phenomenon of health. When we are healthy, everything "flows"; the mood we find ourselves in does not make itself heard or seen. Health could possibly be felt as a form of rhythm in our being-in-the-world connected to time and to the way we are incarnated—to our breathing and to the beating of our hearts (Gadamer, 1993, p. 166). But, health should not be confused with other positive moods like well-being or happiness. Such moods also color our understanding, but in a much more obvious and manifest way than health. Health is a non-apparent attunement, a rhythmic, balancing mood that supports our understanding in a homelike way without calling for our attention.

Let me now return to Heidegger's analysis of the person's being-in-the-world as an attuned understanding played out in the meaning-patterns formed between different tools. I will consider his most famous illustration of such patterns, using the example of a hammer (1986, pp. 69, 84). While engaged in building a house, we do not, in the act of hammering, explicitly focus our attention upon the hammer in an intentional way. We are absorbed in the activity. But then suddenly, the head of the hammer flies off, and the tool can no longer be used for striking nails. The activity is now interrupted, and we are forced to focus upon the hammer, to become conscious of it as a broken tool that must be repaired or replaced if we are to be able to go on building the house. Heidegger's point here is that through breakdowns in the activity, it becomes possible to grasp the hammer as a piece of equipment—a tool—and not just as a thing (1986, p. 73 ff.).

Heidegger in *Sein und Zeit* to some degree neglects highlighting the part played by the body in activity and understanding. The lived body (*Leib*) in *Sein und Zeit* is only a part of Dasein's spatiality (1986, p. 108). It is not identical to the biological body (*Körper*), but Heidegger never extensively discusses the body as a set of tools or as part of the meaning-structure of the world. But no valid argument exists for restricting the concept of tool as it is used in *Sein und Zeit* to things outside the human body.

To place the limit of the meaning-patterns of the world at the surface of the biological organism would be to proceed from a materialist perspective, not from the phenomenological point of view which Heidegger adopts in this work. Once the positions of dualism and reductionism have been given up in the phenomenological attitude, they cannot be brought in again in order to draw a line between the hammer and the hand in the phenomenological analysis. If the hammer—and not the hand—can be said to constitute a tool in the phenomenological sense, this line of demarcation must be drawn from the phenomenological position.

I will now try to show how we can draw this line—not at the surface of the biological organism, and not by denying the hand the position of a piece of equipment—but by stressing the distinction between self and world. If the hand breaks instead of the hammer, the activity will likewise come to an end. The hand, consequently, is also a sort of tool for activity and understanding, constituting a nodal point in the meaning-pattern of the world. The hand, like the hammer, can be repaired or perhaps even replaced by a prosthesis that would adequately perform the functions of a hand. But the situation of the broken hand is also different from the situation of the broken hammer since it is differently attuned. The confusion and annoyance of the broken hammer can hardly be compared to the pain we experience having broken a wrist.

This pain, if continuous, is typically an attunement of illness. The attunement of illness is linked to a defective understanding, since the meaning-patterns of the world are disturbed as a result—in this case—of the body-tools breaking down. The understanding is limited, since a vital part of the totality of relevance has been removed. Still, in a way, it would be wrong to call the body parts tools since they are also part of *Dasein* as a self. They are not only a part of the totality of tools, but also, as lived (*leibliche*), they belong to the projective power of the self.

The hand and the hammer are thus rightly regarded as different forms of tools; but the line of demarcation is not drawn with reference to the surface of the biological organism. This line is, instead, determined through an appeal to the importance the tool plays in the totality of relevance for the human being in question. If the tool belongs to the region we would identify with the self and not with the world in the person's being-in-the-world, then it would consequently be more likely to result in illness if broken.

The self must, however, always be understood as acquiring its meaning and identity by way of its being-in-the-world. The loss of a beloved could lead to an attunement of illness if he or she formed an irreplaceable part of the subject's self as a being-in-the-world. However, the loss of some body parts—such as the appendix—could take place without us afterwards experiencing any pain or illness, and could even mean a recovery from illness.

We might here begin to grasp a possible distinction between handicap and illness. The loss or breakdown of essential body parts belonging to the self typically results in an unhomelike being-in-the-world for the person in question. But in the case of handicap—a person losing his or her sight, hearing, or legs—it might be possible with the help of time, good assistance, and a strong will to reinterpret and change one's being-in-the-world in a way that reestablishes homelikeness.

This "reinterpretation" is a cognitive process that also involves adjustments in the person's attunement and embodiment. It could take a cultivation of alternative faculties of sense—such as touch, for a blind person—or the "extension" of the biological body through a blind stick or a wheelchair. This readjustment might not be possible in every case; unhomelike being-in-the-world will still be the fate for many handicapped. But given the right opportunities as well as intensive and time-consuming training, it remains a viable alternative for many handicapped. The existence of "communities"—groups of handicapped that cultivate lifestyles molded around members' special predicaments—tends to reinforce the possibilities of establishing homelike being-in-the-world, although the meaning-patterns of these lifestyles are different than the standard ones.

Our embodiment forms a vital part of our being-in-the-world. It would appear appropriate to accord the lived body the status of a fourth existential of our being-in-the-world, alongside the other existentials of understanding, attunement, and language. Heidegger in his late work, *Zollikoner Seminare*, came to see this more clearly than he did in *Sein und Zeit*:

> Everything that we refer to as our lived body, including the most minute muscle fibre and the most imperceptible hormone molecule, belongs essentially to our mode of existence. This body is consequently not to be understood as lifeless matter, but is part of that domain that cannot be

> objectified or seen, a being able to encounter significance, which our entire being-there consists in. This lived body forms itself in a way appropriate for using the lifeless and living material objects that it encounters. In contrast to a tool the living domains of existence cannot, however, be released from the human being. They cannot be stored separately in a tool-box. Rather they remain pervaded by human being, kept in a human being, belonging to a human being, as long as he or she lives (1994, p. 293).

That the body is pervaded by human being must mean that it is attuned and involved in the world. As Merleau-Ponty writes in *The Phenomenology of Perception*, the body "understands" and "inhabits" the world (1962, p. 139). Heidegger expresses this notion through the neologism "Das Leiben des Leibes" (1994, p. 113). Existence is a "bodying forth" in the meaning-structures of the world. These meaning-structures are consequently not confined to language or consciousness; as early as in *Sein und Zeit*, Heidegger had given priority to everyday actions in his analysis.

The unhomelikeness of illness would never consist in one or several of the existentials being totally absent. This is not even a possibility, since the existentials are not added to each other as things in the world, but permeate each other in an unravelable way, forming the necessary structure that makes being-in-the-world possible at all. The entire being-in-the-world of the person is always involved in illness as well as in health. This is most evident in the case of some handicaps. Even if a person cannot speak (dumbness) or possesses no "language" in the usual sense (severe disturbance in mental development), one's body would still "express" things in its bodying forth in the world. Analogously, tetraplegics who are able to speak or express themselves through writing (perhaps through some sort of computerized device) would still in some restricted sense be present in their own bodies. Consequently, no handicap and no illness is entirely somatic or entirely mental.

5. Homelike Being-in-the-World

I would now like to address a question that might have been on the reader's mind for some time. It appears that unhomelikeness is a wider conception than illness, since external circumstances may sometimes render our being-in-the-world unhomelike, in ways that we would not choose to refer to as cases of ill health. To be locked up in prison for years and be exposed to horrible conditions is such an example of unhomelikeness without illness.

What should first be stressed here is that the being-in-the-world of the person is what counts as homelike or unhomelike in the phenomenological theory. To live in an environment means to interpret it and assign it meaning through feelings, thoughts, and actions. Thus the life-world of phenomenology is not identical to physical surroundings, but is a meaning-pattern partly created by the person him- or herself. Whether being locked up in prison results in unhomelike being-in-the-world or not depends partly on the world that the person is thrown into (its factual and cultural characteristics) and partly on the way he or she projects this given world of necessities and possibilities in his or her life. The prisoner might in some cases be able to adjust to a homelike existence behind the bars, although the conditions of imprisonment in most cases would offer too much resistance to allow this homelike reinterpretation of the subject's life project.

Second, in my view, the unhomelike being-in-the-world of illness is characterized by a fatal change in the meaning-structures, of the world and of the self. Although self and world are always interconnected in a synthetic way through the being-in-the-world of the self, we can still possibly make a distinction between the person and the world he or she inhabits. We can distinguish between a more general homelessness of being-in-the-world which is due to breakdowns in the "external" tool-pattern, and the unhomelikeness of illness, which is always accompanied by a fatal change in the meaning-structure of the self. As we can easily imagine, the parts of the meaning-pattern that are hardest to replace or reinterpret are the ones tied to our own embodiment. The lived body forms the center of the self. The "body-tools" are typically the most fundamental tools in our life, while tools outside of and apart from the body often are replaceable to a greater extent.

The self, however, is not identical with the lived body, but also has other "mental" characteristics, which are made lucid by the existentials of understanding and language. This distinction between breakdowns in the meaning-patterns of self versus those of world should be sufficient to counter the argument that the characterization of illness as unhomelike being-in-the-world threatens to encompass too many conditions that we would normally refer to as unhappiness rather than illness.

This distinction can also help us to understand the possibility of handicap without illness—that is, the cases of handicapped persons who enjoy a homelike being-in-the-world. These handicapped have managed to make adjustments in the meaning-patterns of their selves in a way that compensates for the maladies they are suffering from. In other words, they have managed to make their being-in-the-world homelike again—given that their handicap is caused by an impairment or injury, and not by a congenital defect—by way of changing their outer circumstances and their understanding of themselves in this altered situation. (In the case of congenital defect, this possible adjustment to homelike being-in-the-world would take place through a process similar to the way that all normal children make themselves at home in the world in developing a coherent self.) In chronic illness, in which, in most cases, the subject suffers from an incurable disease, this reparation and reinterpretation of body and self is much harder, since the disease is an ongoing destructive process, not an unchanging biological state.

The prospects of adjusting to a homelike being-in-the-world through a reinterpretation of the self and its situation in life depends partly on the person afflicted by the malady, and partly on how severe the malady in question is. Some impairments or defects might be so severe that a homelike reinterpretation of life is not possible, while some diseases may be mild enough to allow for a homelike existence. The point of phenomenological theory is precisely to find a starting point in the person and his or her life-world circumstances—not in his or her biology—when it comes to questions of health and illness. This approach, however, is not meant to exclude the importance of biology. The physiology of the body afflicts and sets limits on the different ways we are able to experience and interpret our being-in-the-world.

The homelikeness—and that is, in the phenomenological theory, the health—of some handicapped persons helps us to discern different ways of promoting homelikeness and health for people suffering from different forms of unhomelikeness in their being-in-the-world. Biological therapy represents one significant way of doing so, since the curing of diseases tends to eradicate the unhomelikeness of illness directly in many cases; think of an appendectomy, for instance. Biological technology and therapy are also quite helpful for handicapped people who suffer from biological defects, injuries, or impairments rather than diseases.

Rehabilitation—focusing on changing the self-understanding of patients, improving their mental and physical abilities, and making adjustments in their life environments—forms the next step in the treatment of, for instance, a stroke. But the issue of homelikeness for the handicapped also carries a social and political dimension. To establish homelikeness requires resources for building extraordinary environments in which the handicapped can adjust for their functional disabilities. And it requires respect for nonstandard lifestyles that can be just as homelike as, but nevertheless different from, ours; think of the deaf culture, for instance.

6. Concluding Remarks

The main threat to homelikeness for many people in our Western society today—whether currently referred to as handicapped or not—apparently comes from the streamlining of our culture into a market society where some values, such as speed, efficiency, and productivity, have to be lived up to by every individual. Those who do not are forced into the unhomelike existence of unemployment, uselessness, and poverty. The phenomenological alternative for a theory of health would not deem these people unhealthy, since their lack of homelikeness depends on changes in outer circumstances instead of changes in the structures of their selves.

The tenable path from unhomelike to homelike being-in-the-world in many cases could come through the changing of cultural norms rather than through biological therapy. The mapping and possible correction of biological defects are carried out on the basis of a cultural pattern of normality that is shrinking every day and therefore excludes the possibility of a wide spectra of life-patterns. These life-patterns are now looked upon as being caused by biological malfunctioning rather than being dependent on external circumstances and norms.

The unhomelikeness of many handicapped—whether it is primarily caused by a biological malady or not—could be "cured" by devoting more time and resources to their living environment and trying to change attitudes toward people who may not be biologically normal, but are still human beings like the rest of us. This is a mission for politicians as well as for doctors—especially if we want to retain and cultivate the human aspects of health care and our own society.

Part Four

HEALTH AND HEALTH PROMOTION FROM THE POINT OF VIEW OF FAMILY MEDICINE, MEDICAL GENETICS AND THE CARE OF THE ELDERLY

Eight

HEALTH AND DIALOGUE-BASED MEDICINE

Olle Hellström

1. Introduction

The clinical attitude that I call dialogue-based medicine grounds itself in daily self-reflection on my family practice. It is inspired by the experience that encountering primary health care patients requires a knowledge that goes beyond biomedical theory. I estimate that less than 50 percent of my patients present problems fully understandable by means of biomedical knowledge (Hellström, 1993). Thomasine Kushner (1981) finds that 80 percent of the patients in family practice present problems that require doctors who are prepared to cooperate with the patient to understand and manage personal and existential issues.

Here I will discuss the importance of a dialogical attitude in family practice, where patients' presented experiences often are vague and indefinable. I will sketch and elaborate a radical alternative to the biomedical perspective, discussing different concepts of health and the relationships between the concepts of illness and disease (Boorse, 1981; Hellström, 1999; Nordenfelt, 1993b; Pörn, 1993).

Dialogue-based medicine, as contrasted with diagnosis-based medicine, provides an attitude for meeting patients and for supporting them to consider their illness-experiences. To encounter patients by means of a dialogical attitude, doctors need to reflect on the quality of their own contribution to the dialogue. Such a dynamic suggests a new medical attitude; a conventional and mechanically oriented attitude risks influencing patients to present their illnesses in an impersonal way. We can understand patients better if we include our attitude to them and to their presented sufferings in our main clinical hypothesis.

The practice of a clinical attitude demands reflecting on the evolution of the human being as a symbol-using creature. How, for instance, do we understand each other by means of a language? The most fundamental and crucial part of human communication is its immaterial, invisible, and often unwittingly conveyed aspect, which is carried by verbal communication and other actions.

Representatives of different medical subspecialties argue that the clinical goal is to maintain and practice a holistic attitude. Their arguments are often based on an assumption that holism is a matter of quantity: The more aspects included, the more holistic an attitude. I suggest that we are practicing the most holistic clinical attitude when we keep being aware of our technically dominated education and its impact on the patients' way of presenting their illness-experiences.

I reject the view that physicians skilled in performing well structured talks with patients, that is putting exhaustive and directed questions to them, also produce better and quicker diagnoses. This view is not relevant for the meaning I ascribe to the phenomenon of dialogue. The concept of dialogue medicine denotes an attitude aimed at supporting the exchange of meaning between doctor and patient. This helps both parties to deepen the understanding also of the non-ostensive aspects of the experienced illness. Furthermore, increased competence in dialogue medicine techniques in general practice will lead to optimal use of other specialists' services; to reduced medicalization of human non-medical problems expressed as bodily disorders; to increased precision in diagnosis; and to improved capability to meet and manage patients with poorly defined ailments.

Finally, I discuss the concept of diagnosis-oriented medicine and its connections to dialogue-based medicine, seen as an overarching clinical attitude where diagnostic efforts are regarded as useful complements. This way of seeing things makes it natural to start all encounters by openly inviting the patient into a dialogue. When the options of the dialogue are exhausted, I complete the meeting by focusing on possible bodily faults. In other words, I use my personal capability to understand the patient as an expressive human being in order to assist him or her in finding a less illness-oriented stance. We consider the existence of experienced bodily disorder together. This attitude is indispensable, especially when patients experience themselves as ill, yet no diseases are to be found.

During the 1980s, many long-lasting dialogue sessions with "Paula" (Hellström, 1998) challenged me to improve my practical dialogue competence. She was convinced that she had cancer. Her non-ostensive problem was an existential dilemma. Her desired and unavoidable divorce was connected to the moral issue of having to abandon a young son. My dialogues with Paula helped me to elaborate a useful structure in my clinical practice and encouraged me to bracket my biomedically conditioned thoughts and actions until they were appropriate to apply.

2. Health

According to the health theories of Lennart Nordenfelt (1993b) and Ingmar Pörn (1993), an individual is in a state of good health if, in a given environment, he or she has the capability to reach his or her vital goal(s). Their equilibrium theories of health view the human being as an acting subject. They consider ill health as an imbalance between the acting individual's repertoire, vital goals, and environment.

Christopher Boorse's (1981) health theory relies on the biomedical conceptualization of the human being. Statistics are also a crucial aspect in his theory. It positively correlates health to the normality of detailed measurable parameters of the individual seen as an organic system. Fundamental differences exist between the two types of health theories. Nordenfelt and Pörn begin by defining health, and their theories comprise no strong correlation between disease and poor health, whereas Boorse takes his starting point from the concept of disease, defines it in detail, and conceives of health as the absence of disease.

A health theory that draws on natural science and describes good health as having no measurable deviance from what is statistically normal has little to offer doctors who see patients as subjective human beings whose experienced symptoms can refer to something beyond statistically measurable disorders. Human experiences lose their essence when, for a theory to maintain its consistency, they are reduced to impersonal behavioral movements. When doctors see health as a state of equilibrium between a person's capability, their vital goal, and their environment, they can also integrate the theory into their attitude toward patients whose illness-experiences do not refer to physical disorders. One central vital goal is to establish a comprehensible, manageable relationship to a world of meaning where people can understand each other and make themselves understood.

I suggest that some people can try to maintain their sense of good health—that is, uphold their equilibrium between their capability, set vital goal, and environment—by believing that they have been afflicted with a disease. This image helps such patients to remain unaware of the threatening sense of responsibility for a feared inability to reach their most vital goals. If a patient presents symptoms of an experienced disease and the doctor takes the expected medical steps to identify and cure his or her "disease," the patient can still feel harmony between his or her capability, vital goal (to have exposed meaningful actions understood and confirmed), and environment. Patients can also unwittingly redirect the illness-based search for meaning toward long-lasting efforts to convince doctors that they do have a disease. In such a case, doctors have use for a health theory that provides them with the basic prerequisites for distinguishing between the effects of an existing disease and an illness-experience where no disease exists.

3. A Report of a Patient Encounter

I will here give an account of some theoretical and practical aspects of the dialogue attitude as applied in the case of a 60-year-old male patient, "Heikki," who has been misunderstood by doctors and therapists for 30 years.

"How do you know when a patient is to be met as if some bodily disorder causes his or her problem or when presented illness-experiences are to be seen as a personal, non-physical issue that has found expression in a bodily disorder?" So asks a young colleague when we are discussing the usefulness of clinical dialogue, and Heikki's experienced illness is on our agenda. Little evidence indicates that Heikki suffers from a disease. Many doctors have examined and treated him as if he had a bodily disorder to be identified and treated. During our seven encounters, he gradually reorients himself and improves his health.

Heikki was born in Finland in 1942. He moved to Sweden in 1970, leaving his wife and a daughter behind. Soon Heikki found a new wife and became the father of two girls, now 27 and 22 years of age. This new relationship dissolved after 12 years of marriage, in 1988. Heikki has now married for a third time.

Heikki's biggest problem is an uncomfortable sensation in his head and a painful, weak body. He experiences this discomfort periodically and finds it extremely bothersome. These symptoms began when he divorced his first wife 30 years ago, starting in his head and radiating to the rest of his body. His discomfort is always present in its basic form and increases in waves, culminating in intolerable pain.

Heikki's affliction has given rise to an ear pain that makes him indifferent and prevents him from being happy. After many years of baffling efforts to define the cause of his pain, Heikki was going to get an operation. But something gave way in his ear, "… an unbelievable lot of pus welled forth," and the operation was canceled. The ear specialist could not explain the cause of the symptoms. A week later, Heikki had an operation.

During our seven encounters in Heikki's home, he always repeats what he has gone through as a patient. No doctor has given him a trustworthy causal explanation of his experienced affliction. His files show that he has gone through many medical and technical examinations of different organs and that psychodynamic, cognitive behavior and psychiatric therapists, acupuncturists, and Qi Gong therapists have treated him. Therapists and physicians have used many different psycho-pharmacological remedies to treat him.

During our dialogues, I realize how strongly involved Heikki is in his body and its many experienced dysfunctions, and how skilled he has become in describing his bodily sensations. Heikki believes that he is afflicted with an infection in his head. One of his most overt goals is to get a technically comprehensible causal explanation. He assures to me that his trouble has nothing to do with his psyche.

After having listened to Heikki for about three hours during our first two sessions, I carefully started to inquire about the most conspicuous aspects of his illness narrative. Experienced physical afflictions, weakness, and sadness are the themes he expressed most clearly. I also recognize my sense of incapability in the face of Heikki's intense, anguished, and eager efforts to make himself understood. Heikki's conviction of having been afflicted with a disease appears to arrest him. His ideas partly derive from and are strengthened by his meetings with doctors who for years have subjected him to innumerable lengthy fault-detecting procedures. Unable to find the faults searched for, his doctors used to suggest that Heikki's psyche was to blame. Without this constructed theoretical temporary solution, they would have had to admit their helplessness.

Heikki still searches for an explanation of his 30 years of suffering. In our encounters, he tries to take command by telling me about all his inexplicable bodily experiences. Body language, gaze, and gestures support his impassioned accounts. Rationalizing traits are common in his narrative. For instance, he feels that he is no longer capable of reading technical blueprints, which he believes limits his usefulness as a worker. When Heikki's superiors offer him a new and less challenging task at work, he knows that his incompetence depends on his illness. He also believes that his work colleagues avoid him and look upon him as an inferior worker. "They do not know that I am afflicted with a disease," he tells me. His ideas about general and increasing hostility to immigrants in Sweden help him to cope with feeling lonesome at work.

Heikki prefers to talk about and live in the past. He chooses historical books. In many respects, he demonstrates that he avoids facing his future. When asked to talk about his plans and desires, Heikki reacts by telling me with intensity about all the illnesses that have stopped him from realizing any dreams. When I ask him what he will be doing when I leave, he replies that he will go back and work in his garden: "Today my body is fine and will not stop me, but it usually makes me unable to do what I want to. This keeps me isolated and lonely."

Heikki used to be an enthusiastic motorcyclist. He has been looking at different scooters lately, but he is not at all excited about the thought of buying one. "It" (his illness) strangles his joy. He describes a recent trip abroad as a complete disaster. Though his illness does not necessarily stop him from doing what he wants to, his suffering takes his joy away and makes him indifferent. He does not buy any clothes and does not even want his sexual incapability to be cured. He also finds his intellectual potential reduced by "it," and again he refers to his decreased ability to read blueprints.

4. Narratives, Diagnostics, and the Usefulness of Dialogue-Based Medicine

Exchange of information, counseling, supervision, and patient education are central concepts in medical literature on doctor-patient relationship consultation (Byrne. 1984; McWhinney, 1997; Tate 1995). The concept of dialogue-based medicine is especially useful for negotiating with patients to encourage them to reconsider their views of themselves and their resources. Dialogue medicine concerns an attitude to patients in clinics, not a method or a technique for treating them, and should not replace diagnostics. Accordingly, I will touch upon the goal of diagnosis in the patient encounter. One aim of incorporating dialogue into the clinical patient meeting is to create a firm ground for a more precise diagnosis.

With no claims of making an exhaustive account, I will first discuss possible ways of meeting patients and their experienced diseases. Then I will sketch a model for a dialogue-based clinical attitude, a fairly new way of approaching suffering people in the context of medical care.

Heikki represents a resource-demanding category of patients who are quite common in primary health care. His fate, while challenging, is useful for investigating the competence of general practitioners in the skill of professional dialogue. Heikki's files reveal how different doctors and therapists have perceived illness. To biomedically minded general practitioners, a presented symptom indicates the probable existence of some bodily disorder. They believe that they must identify the disorder by making a diagnosis, use corresponding action programs, and eliminate the causes. If their plan is not successful, they turn to the second-best alternative and treat the symptoms. As a third option, they can consult a specialist who has advanced technical equipment for identifying or excluding the diseases in mind. A less obvious motive for a referral can be a wish to avoid further responsibility for the patient, whose biomedically incomprehensible symptoms give rise to a feeling of uneasiness and threaten the doctor's professional self-esteem.

A mechanical view of the relationship between body and mind is guiding psychosomatically oriented general practitioners. By means of a linear apprehension of time sequences, they look for something in the psyche that can make an impact on the body, which in turn triggers, for instance, pain signals to the brain. This explains the presence of pain without a bodily disorder. Modern pain theorists claim that chronic, untreated pain can generate new central nerve connections (Coderre, 1993; Harris, 1999) that maintain the sensation of pain even when the original source of pain is no longer active. If pain is not treated early and aggressively it can turn chronic—an attractive theory for the pain pharmaceutical industry.

Orthodox Freudian therapy, to mention still another objectively oriented tradition, is guided by the hypothesis that the patient's personal history conceals the clue to understanding the symptoms. Something in our past history, in childhood, or even still earlier has made us who we are. History-mapping is hard work. We all try to resist unveiling what we do not dare to become reminded of.

Whereas psychotherapists search for a code that helps them to interpret the patient's ostensive symptoms, existential psychoanalysts with a phenomenological approach are more open to the non-ostensive aspects of what is expressed in a meeting (Binswanger, 1958; Boss, 1979; May, 1983, Yalom 1980). An existential perspective is a good start in every encounter. It helps doctors to identify the preferable level on which to meet the patient's presented problem. Reflecting on non-ostensive aspects in a meeting, for instance, and on patients' search for meaning can contribute to resolving the images that find expression in an experienced bodily disorder—for instance, pain, vertigo, or other afflictions.

Rollo May (1983) holds that the main task of a therapist is to try to understand the patient as a human being in his or her world and that other human issues should be subordinated to this understanding. Existing persons, in May's theory, are centered in themselves. A threat to the person's center is also directed at their existence or being. What appears to be a symptom may be the patient's expressed drive to defend his or her being, even if the experience presents as a bodily disorder. We should understand the being as the existence through which we gain the option to become the people we aim to be. Every existing person is characterized by a will to self-confirmation and a need to retain and protect his or her center. Those who have entered a self-created reality dominated by a disease experience and who dare to talk about that which they disguise by means of an illness risk losing their remaining contracted living space.

Patients need courage. Without courage, one will lose one's being, says Paul Tillich (1952). When patients mobilize courage to approach their non-ostensive dilemma, when they put their existence or being at stake, they can use their human power for reaching more constructive goals than stopping the unthinkable from happening. An existent person is able to transcend himself or herself and enter into a community with others. This, too, is dangerous. Patients are sometimes afraid of losing their center. They therefore force themselves to live isolated in a continually diminishing world. A shared feeling of having been afflicted with pain, for example, can also give rise to a sense of togetherness, a fellowship binding all those who are in pain.

Heikki has a "reason" for being confused: It helps him to stay unaware of what would otherwise bring him into contact with his responsibility for not being on the path that will lead him to be or become the person he truly is. He keeps nurturing the idea that he is exposed to or afflicted with something, and this idea contributes to the confusion. Doctors have for years unwittingly encouraged him to maintain his illness-experiences. To become aware of his ambivalence, Heikki risks having to give up a high personal priority, a significant aspect of his self-created center. He protects his consciousness from such thoughts by occupying it with an illness-experience.

How do I best help Heikki to illuminate his being-in-the-world? I let his personal context guide my attitude and I try to understand him as he is at present. This psychological dynamic makes sense in the existence where Heikki lives his immediate life. Existential therapy stresses presence, which means that all parties perceive the relationship between therapist and patient as real. Being present must not be confused with a sentimental attitude. To be present in a genuine encounter, I cannot think in terms of theories, historical explanations, and diagnoses.

Others presume me to be an expert, but if I do not use myself as a human being, my expertise will be irrelevant and even harmful. As an existential person, I may even be inclined to avoid the anguish and the uneasiness that Heikki's dilemma awakens. A technical view of him may be my most convenient personal anguish-reducing means. Possible ways out of an anguish-provoking encounter are to refer the patient to a specialist, to make a diagnosis, or to use other classifying methods.

5. Explaining Objectified Patients or Understanding Unique Human Subjects

My dialogues with Heikki give rise to many questions. How can an intellectually well-equipped man who has worked full-time for more than 30 years hold on to the same illness-image, one dealing with cause and effect, and let himself be repeatedly exposed to apparently useless examinations and therapeutic efforts? How can doctors, claiming that their actions are scientific and evidence-based, hold on to the same physical and technical action program when, judging from experiences and authentic files, they again and again miss their professional target? Does a symbiotic connection between doctor and patient exist that is deeply rooted in the irrational belief in natural sciences and diagnostics?

Do doctors feel anguish at the thought of having to reconsider their biomedical basis of knowledge? Do physically sound patients who experience themselves as ill fear leaving an illness-oriented self-image that makes them feel reasonably coherent? Is the body a safe retreat from a confusing external world? Do patients unwittingly rely on a mechanistic action perspective of the health care system to escape from a world that they find largely unfamiliar? Why do Heikki and many other patients so intensely dislike doctors' implications that the cause of an experienced ailment may be a psychic disorder?

6. Three Life Modalities

Psychosomatic researchers have tried hard to demonstrate theoretically and experimentally that the psychosocial environment can make people develop a disease. However, research such as experimental pain studies usually overlooks the first step in the cause-effect chain of the individual pain experience. By calling the pain experience "pain behavior," researchers can examine pain in rats, for example, ignoring that the world of animals is not interpretable in human terms.

Animals live exclusively in a physical environment (in German: *die Umwelt*). The human world consists of a physical environment as well as a commonly shared world of meaning (*die Mitwelt*) and an intra-personal self-world (*die Eigenwelt*) (Binswanger, 1958; May, 1983). Only experiences, not behavior, give us access to those two life modalities, and human beings cannot do without them.

When encountering patients with ostensibly simple ailments, we need to realize that human beings live in three life spheres or modalities: in relation to themselves, to the world of meaning shared with others, and to a physical world. We do not need to rely on theoretical temporary solutions (such as psychosomatics) when facing biomedically confusing symptom presentations. Patients always have access to their body, both when they have words for what they experience and need to convey and when they lack such words. Without their body they would not be able to express themselves verbally, relate to themselves, or consider their own thoughts and words. My words mean something to me only if a have a picture of myself as the person who says them. Words and meaningful actions help us to feel confirmed as persons who take part in a shared world of meaning.

Patients can present their experienced illness to doctors and can establish themselves as people who are able to make others respond to them as human beings. The responses indicate the existence of a "common communicative environment" (Toombs, 1992). Patients with a long-standing pain-experience who eventually are diagnosed with fibromyalgia, for instance, become happy although the diagnosis denotes lifelong pain and fatigue (Hellström, 1999). The explanation for this apparent paradox may be that any diagnosis that gives structure to the patient's existence is welcome. But do not these and other similar observations imply the absence of more sound alternative views of health, diagnosis, illness, and disease?

The biomedical scientific culture dominates doctors' views of relevant clinical concepts. Science expects us to ignore that the phenomenon of experience is our only access to the human world, a constantly present and insoluble coherent world of self, meaning, and physical relations. We live in an era and in a culture where we usually ask "Why?" when we want to know the meaning of something and we unreflectively structure the answer in terms of cause and effect. We search for fragments and factors that can give us a firm basis for clarifying things. When patients sense something in the body, they ask for medical technical explanations. Patients try to maximize the possibility of making themselves understood by presenting themselves in a form that they expect biomedically educated doctors to understand.

Theoretically, doctors can put other questions to patients than the kind that aim at explanation. One such question is "What for?" Guided by "What for?" questions, we approach a radically different human life dimension that is connected to intention, context, and personal choices. The intention of an action is personal and searches for a personal response. The observation that patients unwillingly admit the presence of intentions in their exposed illness-experiences could be useful in medical practice. Patients can escape from their personal responsibility by remaining unaware of the intentional aspect of their illnesses. Such a strategy allows them to manage the threat to their sense of meaning, comprehensibility, and manageability—the crucial three aspects of coherence in Aaron Antonovsky's theory of health (Antonovsky, 1982).

Can we drive ourselves to be ill to remain healthy? For what reason, for instance, does an actor sense abnormal gut movements and fast heartbeats in connection with the thought of an approaching performance? Is the experienced uneasiness even purposeful, looked upon as a bodily message to the actor to reconsider his or her view of the audience? The approaching performance as such does not, at present, affect the actor. He or she actively relates and ascribes a self-created view to the situation.

The question I suggest is not "why" this happens, but "what for," the meaning behind it. The actor does not face the effect of some stimuli from the psychosocial environment, but that environment does expose a self-created image of what is to come. The sensed fear of the unknown future event expresses and has a connection to a lack of words and thematic meaning (Bullington, 1999). The actor does not have access to all the appropriate symbols of what is to come. At the same time, he or she is bound to communicate not to lose contact with a world of meaning shared with others. The actor can manage a threatening existential dilemma by searching for meaning within it and by ascribing familiar meaning to it. Sensed anguish could be part of the actor's threatening feeling of being alone, misunderstood, an outsider who must confront human emptiness, and so on.

Human beings can manage the anguish of nothingness by transforming it into fear of the experienced bodily "disorder." In that case, all three life modalities are involved. A doctor to whom our actor perhaps presents his or her bodily experience is free to subject it to mechanistically oriented examinations, or to cooperative dialogue. The doctor can try to classify and eliminate what he or she perceives as diagnosable symptoms or focus upon the presented experiences as existentially interpretable messages and try to understand them together with the patient on the most equal terms possible.

7. Heikki, I, and Our Three Shared World Aspects

For thousands of years, human beings have had the ability to handle symbols in communication with each other. Developing this communicative competence is an essential part of our maturing. Symbols play a profound role in our existence. We incessantly want others to confirm us as members in a world of symbol-creating and symbol-using human beings. The only means at hand are verbal or other types of meaningful actions. When others understand and respond to us, we feel that we are part of the shared community we have been searching for. We do not exchange any kind of material substance: our world of meaning is immaterial.

Heikki experiences his illness, but he may or may not have a disease. Let us presume that something other than a disease sheds light on his illness experience. What then makes him feel ill? How could he feel like having a disease when he does not have one?

Heikki exposes his self-created picture. He holds onto an idea that doctors will understand him in his way and that they will be able to help him if he manages to present his self-picture in a convincing way. Therefore, he tries to convey his experiences of affliction by means of a symbol language which he hopes will be intelligible.

If, hypothetically, Heikki has problems not with his disordered body, but with himself as a citizen in "*die Eigenwelt*" and in his relationship to others in "*die Mitwelt*," then he has to turn to "*die Umwelt*"—unless he has access to the appropriate symbol language. He therefore expresses not himself, but his body, symbolically and in a fairly primitive way.

In our encounter, I recognize Heikki's capability to use his body in order to expose his relation to himself. I try to meet Heikki, not only as a faulty organism with a sound-producing faculty, but as a person whose meaningfully conveyed bodily experiences indicate efforts to manage an existential dilemma.

What I say about Heikki as a meaning-creating and meaning-conveying being also counts for doctors. We, too, want to feel that we take part in a "common communicative environment" (Toombs, 1992). When we encounter a person whose narrative is hard to understand, we tend to ascribe meaning to it so as to make us feel less professionally confused. We often start from an *a priori* image of the patient's presented problem and look for details, selecting mainly those factors that suit our initially created view—a result of our biomedical education. Corresponding action-repertoires say little about the importance of being open to, listening to, and seeing patients' desires, demands, ambivalent messages, and experiences. A tacit demarcation line between subject and object is the primary obstacle to constructive dialogical encounters. When this theoretical hindrance is overcome and the fragmented human being is healed again, an attitude that allows us to understand patients as persons before we identify different potential causes and apply diagnostic and therapeutic scientific means can flourish (Hellström, 1999).

If we see Heikki's intentions primarily as a question of maintaining a sense of coherence and meaning, we also see his reasons for presenting himself in the way he does. He wants to be understood, not explained. He tends to protect his retreated and defensive mode of relating to his world of meaning. He knows what he has achieved, but not what he might gain.

In the past, when Heikki presented his choice of direction to doctors, he expected them to use a map similar to his. He picked a "co-driver" who would confirm his already-made choice of direction. He was not prepared to listen to and be open to a dialogue partner who tended to threaten his body-based notion of the kind of suffering he had experienced. In his earlier meetings with doctors, Heikki got indirect, vast, and relieving confirmation that his experienced bodily disorders were "real" and had nothing to do with his personal responsibility.

8. Who Is Heikki?

Not knowing how, why, or what for, Heikki immediately heads for his mother's breast during the first moments after his birth. He is able to make the most direct eye contact with the area between the breast nipple and his mother's eyes. Her gaze provides an initial safe relation from which base he gradually can widen his excursive loops from the mother and into the world (Bowlby, 1969). Heikki uses the immaterial option to start creating his life-world. Biological instincts still govern him and he is completely dependent on the already existing world into which he has been thrown without asking for it.

When his empty stomach screams, he all of a sudden, like a miracle, has a breast nipple in his mouth. He then feels his mother's skin, she talks to him and responds verbally and bodily to his instinctively produced sounds. In this way, Heikki learns the appropriate meaning of used gestures and words (Winnicott, 1984). Bodily motions become actions saturated with meaning. Heikki also learns how to recognize different bodily sensations and to communicate that he is hungry, freezing, and, later, that he wants to use the bathroom. This life-long learning leads to an ability to manage the symbol form of many kinds of relationships.

Heikki learns to understand others and to make himself understood by means of immaterial symbols. Unless he acts symbolically in an adequate way when trying to eliminate some experienced threat, his body will make itself heard. As long as his empty stomach is not filled with milk from his mother's breast, he will cry and signal a feeling of emptiness. Later, when Heikki can say, "I am hungry," he has reached a vital and appropriate communicative goal. Another lifelong career in human symbol-managing has just started.

Heikki also encounters situations where he no longer receives the confirmation that he expects. Perhaps something happens that he has no words for, for instance some personal bereavement or loss. If this verbal deficit concerns his personal center, he does not easily know how to expose himself in order to share his dilemma and search for meaning and confirmation. He becomes a little less Heikki, and risks losing his sense of coherence.

To restore the connection with his created world of coherence, in relation to which Heikki again feels at home, he returns to an era in his personal development during which his ability to use symbols was less sophisticated. The primitive sounds produced by his empty stomach used to result in a pleasant feeling of nourished satisfaction. This materially based exchange of bodily reactions is part of a pre-symbolic context. Now Heikki no longer screams to reduce his unpleasant feeling of emptiness. He has developed a capability to reach the Other by means of symbols carried by spoken words.

Heikki can still experience a less differentiated gut feeling like the one that he sensed as an infant when his stomach needed feeding. Now he can talk about it and maintain the bodily sense of it. The corresponding words still have a tacit meaning connected to what was going on when the gastric mucous membrane was red with lack of milk and charged with a symbolic quality by his mother's milk. The symbolic aspect of Heikki's exposed experience of a disorder is hardly understandable without considering also the historical bodily aspects.

9. An Approach without Theoretical Temporary Solutions

My sketch showing how bodily sound patients elaborate, maintain, and expose their illness experiences is a reply to the common medical efforts to manage patients' confusing illness experiences by referring to the concept of psychic, psychosomatic, and similar theoretical constructions. I am looking for a coherent and radically new basis for understanding expressed illness-experiences that respects the idea that body and mind are inseparable entities that cannot affect each other. Health-care providers can find the solution in cooperation with patients by meeting and interpreting patients' ostensive suffering as an aspect of their efforts to make themselves understood. We will help many more patients if we do not try to explain the qualities we ascribe to their illness-experiences in terms of cause and effect, and if we do not let mechanistically oriented theories of health, illness, and disease guide us.

Our ability to differentiate between symptoms and presented illness experiences is vital for a successful practice. Also, patients use the concept of symptom as doctors do: as something that indicates the presence of a detectable disorder, a disease. If we hold on to the notion that we confront an intentionally presented illness experience, then we do not ask how to explain our image of it. We try to understand what we meet; namely, a personally conveyed bodily experience. A small problem exists if the presented experience does reflect a bodily disorder. But if we start the encounter by trying to diagnose something that we believe should be met objectively, unaffected by our personal view of it, then we run the risk of missing the diagnosis and of misunderstanding the patient's subjective context.

A crucial question is whether Heikki's illness has to do with insufficient bodily functions or with his personal qualities as a communicative human being. Does he prefer that we see him and listen to him, or would he rather that we expose him to medical fault-detection? If I do not put these questions to myself, patients like Heikki run the risk of being absorbed in an impersonal medical production line and exposed to unlimited examination procedures.

The symbiosis between a doctor, who has an impact on what his patients express in his office, and the patient, who needs to maintain his or her illness experience, directs the attention of both dialogue partners to the ostensive aspects of the encounter and stops them from approaching less apparent possibilities. When I address Heikki's tacit desire to be nourished as a symbol-using creature, he feels that he can make me understand. He then modulates the picture behind his words and dares to approach what he until then has unwittingly disguised by holding on to a cause-effect view of his illness experience.

10. The Seven Phases of Dialogue-Based Medicine

How do we know when a patient's illness presentation demands that the doctor understand the patient's disguised sense of meaninglessness, and when diagnosis, treatment, and cure should be the most appropriate actions? The answer is that doctors never know and that a dialogical attitude is the best possible way of encountering patients.

The meeting between doctor and patient is a meeting between two persons who have created themselves and their images of the world. Each image derives from relationships to parents, schoolmates, co-workers, and from other essential persons and contexts. The two unwillingly give up their self-created values and attitudes. The ranking of these values decides each individual's personality (Frankl, 1966).

Heikki, for instance, has been creating himself and his view of the world for almost 60 years—the last 30 of which he has cultured and integrated an image of himself as afflicted with a disease. Biomedical and objective efforts to decide whether his experiences, or those of other people who feel ill, depend on bodily changes are fruitless. To apply an objectively guided approach in a meaningful way, we must first decide in which of the three human life modalities we should seek the presented problem.

A. Introduction Phase

In the introduction phase, shaking hands and making eye contact can already guide the dialogue in a successful direction. Heikki sees how I sit down, whether my body language reveals an open attitude, whether my arms are crossed in front of me or they rest in my lap, whether I listen attentively, whether I am involved, and so on. I have as a primary goal to present myself as prepared to meet Heikki on his terms, not mine.

B. Getting to Know Each Other

In phase two, I see to it that I find Heikki where he is, in line with Kierkegaard's well-known thesis. In practice, this means that I initially act and try to say what he expects me to. Meeting Heikki and finding him where he is as a person with all his values means, for instance, to perform some of the bodily examinations usually done in medical practice. I am aware that Heikki wants to be seen and heard as he sees himself. If he wants to be heard but refuses to listen to my answers, it may mean that he is preoccupied in protecting a self-image that he is not prepared to have reconsidered. He might even need his illness and prefer to meet a doctor who just confirms his prejudices.

Such meetings are demanding also for doctors. General practitioners have to receive many emotion-laden, trying messages, and are not inclined to expose their thoughts about patients to have their professional competence confirmed in the encounters. Already, in this phase of the encounter, a clinician skilled in dialogue can see whether theoretically reasonable biomedical examinations are useful.

C. Consensus

During the consensus phase, I carefully tell Heikki how I look upon his experiences, wishes, and demands. I try to make myself understood by suggesting, negotiating, and asking him to deepen the clarity of his conveyed reflections. When I tell him how I see his situation, he will at best sense that he talks to a doctor who wants—and dares—to understand him. Heikki needs courage to reconsider the set of values that constitute him as a person and to throw himself into a completely unfamiliar surrounding. I have a good opportunity to see whether Heikki is prepared to let me enter into his private world of meaning and whether his illness experience indicates the presence of a disease or a need to be understood on a human communicative level. Does he rely on an excuse that demands illness to protect himself from being found responsible for his personal failure?

D. Negotiation

In phase four, I try to use a set of open questions aimed at orienting myself in Heikki's way of seeing himself and his world picture. For example: "If I understand things right, you feel ... (such and such). What would you like me to do for you today?" "You have obviously experienced your ... (pain/ailment/vertigo/tiredness, etc.) for ...(days/months/years). What do you think it can be?" "What have you done to reduce your (pain/ailment/vertigo/tiredness, etc.)?" "How do you think we ought to see your problem in order for us to manage it and for you to re-establish your health?"

Some of these questions have been modified from the Norwegian general practitioner Kirsti Malterud (1992). We should see them as aspects of our ongoing clinical encounter negotiations. I must not use directed questions that confirm my expectations. The questions I ask are intended to show to Heikki that I respect his ability to see and reconstruct his picture of what is going on and that I am prepared to see the matter in his way. I let him know that I am not responsible for finding the connections between such facts as will explain why he is ill. My questions imply that I perceive his ability to reflect on his experiences, not only on mechanistically understood bodily conditions. The questions help me to focus on the meeting as a cooperative enterprise, not as a distanced, object-oriented, instrumental search for bodily failures. We are both creating and guided by our shared goal to understand Heikki's central problem.

Our ongoing encounter will decide the way we look on Heikki's presented experiences. When I respond to his reflections, I imply that his ailment is not serious from a physical point of view, that the referrals to other specialists that he desires can be postponed, and that he does not suffer from psychic problems. Heikki's surprise gives me the opportunity to say that the concept of the psychic is a theoretical construction, a temporary solution often used by doctors not to have to reconsider their main biomedical hypothesis. "This hypothesis," I add, "says that diagnosis is the primary goal in medical practice."

As I hold that Heikki's illness appears to be neither somatic nor psychic, he (as most patients would) then asks what kind of illness he has. My reply is: "Well, I do not know, really. Let's work it out together!" The advantage of such an answer is that Heikki can change his view of his conveyed illness experience from being seen as an object for me to explore into something we can manage as a project for cooperative understanding, a good ground for symmetric dialogue.

A subject-object enterprise has transformed into a subject-subject dialogue. The atmosphere of our dialogue stops Heikki from remaining passive and waiting for me to examine, classify, and manage his sensed bodily disorders. It makes him involved in thinking, listening and talking, and he becomes inspired to convey his thoughts in spoken words—a key means for overcoming the needs of being ill. The dialogue creates a symmetrical basis for reciprocal understanding. It makes fertile ground for a healthy reorientation. Heikki does not manage without my general knowledge and experiences, and I will not succeed without his images or his preparedness to unveil them.

E. Reorientation

In phase five, I attentively and sensibly try to find out whether Heikki is prepared to view his illness experiences in a new and different way. I can, for instance, use my earlier experiences from many other encounters where patients wished to be examined, diagnosed, and cured, and where they eventually understood that they should not see themselves as objects for others' examinations. Such narratives may facilitate Heikki's willingness to reconsider his original reflections in a constructive way and mobilize his self-reflective capability.

F. Genuine Dialogue

In phase six, I apply my readiness to reconsider my attitude in the encounter. The main characteristic of genuine dialogue is the preparedness of both participants to reconsider their initially held prejudices. This is no problem for skilled dialogue-oriented clinicians. They are not afraid of being open in practice, as they know that openness is the crucial aspect of the dialogical attitude. Without this openness, the doctor is restricted to his preconceived picture of the patient's suffering.

Heikki's presented illness experience may represent an effort to stop him from making contact with an unwanted and unthinkable experience. It is likely that he tries to maintain his illness-laden self-picture, an image of having been afflicted with a disease. This thought construction gives him a sense of not being responsible for his unwanted incapability to reach his set life-goal. A skilled dialogue clinician senses the presence of such traits in the patient's responses, which motivates him or her not to further the biomedical investigation process.

G. Consolidation Phase

In the seventh phase, I consolidate my dialogue position by confirming Heikki's new self-image. I try to show that I see his growing ability to make himself understood as a person who needs my assistance without necessarily having been afflicted with a disease. I am especially careful to tell him that I see (on my latest visit to him) his new scooter at the front yard. For a long time, Heikki had been dreaming of buying a scooter, but he hesitated because of his poor health. The brand new, shiny blue scooter is an unambiguous sign of health improvement.

The seven phases of dialogue-based medicine will seldom appear in the order just accounted for, and some phases will be less distinct or appear together with aspects of other phases. Sometimes patients adopt the experiences of new ailments when they have abandoned earlier ones. Seldom can a practitioner recognize all phases during a single encounter. The health-regaining process can stop for years, pivot for a long time in one phase, return to earlier stages, or skip some phases.

11. Concluding Remarks

Dialogue-based medicine rests on the idea that human beings create themselves and that no one else can change another's self-picture. We were all once thrown into a pre-formed world and had to try to manage it and create ourselves. Our mother's affirming responses made us feel rewarded when we adopted some values, so we made them our personal ways of relating to the surrounding world. Our experienced relations settled into us in symbolic form and we can mobilize them when needed. If we must not or cannot voice such integrated images in a situation, we are ready to turn to something that we can voice. This means that we prefer schemes that match our range of allowed, familiar, and homelike images.

Patients who do not dare to see beyond their personal center can give voice to something that does not interfere with their central values. These self-maintaining images must not be so differentiated, discursive or distinctly presented that they can easily be met and reconsidered in a convincing way by doctors. Illness-images are good for keeping unwanted meaning disguised or hidden, prevented from being seen or valued.

This approach requires doctors who can relate to patients in a way that inspires them to reconsider their self-images and associated unnecessary self-perpetuating illness-experiences. Doctors whose knowledge is only biomedical risk supporting these patients' unconscious illness-careers and flights into lifelong incapacity. This potential clinical capability interests me as a doctor: a role that encompasses an ability to balance between seeing the particular and general, the individual and commonly shared, and the ostensive and non-ostensive aspects of patients' presented illness experiences.

Medicine needs theoretical maps that help us to find our way in practice, maps with symbols that do not represent patients' or doctors' perspectives, but mirror our shared, cooperative context. If we restrict our possibilities when trying to find our way in practice through creating new diagnoses, we will not be able to break with the widespread, persistent medicalization of basically non-physical problems.

The proper theoretical perspective can symbolize the human sphere that primarily concerns the relation between doctor and patient. We cannot manage its symbols by means of traditional medical science; they will gradually be woven into us and become personal traits of our professional dialogical attitude. When we can manage our encounters and apply a more rational attitude to existing impersonal map symbols, we will be able to contribute to our patients' hope. They will find their meeting with us meaningful and will dare to convey even the unthinkable. When we, together with patients, can find the words that correspond to their vicarious pain, vertigo, and other symptoms, they will be encouraged to give up their bodily experiences and re-establish contact with their world of meaning.

If we do not want to rely upon, for instance, applying psychosomatic explanations to human conditions that are too confusing to understand by means of natural science, we can recall that our body is prepared for expressing what we have and do not have words for. We would not be able to nuance ourselves if we could not relate to ourselves. A word means something to me only when I can see myself expressing it. I use words and hope to make others confirm me as the person I am busy creating.

Encounter reports are the best raw material for creating the knowledge that makes dialogue-based medicine successful. Dialogue competence concerns a field of knowledge that does not recognize the border between subject and object, and that belongs to neither the subjective nor the objective thought-sphere. The encounter reports can work as classical myths in their search for reality, and as analogue models that help us to catch sight of and understand what we already know and unreflectively relate to.

Those who have read and tried to understand the narrative of Heikki and I will, perhaps, no longer automatically prioritize fault-finding in their encounters with patients who present themselves as he did. Heikki and many of my other patients have encouraged me to create an appropriate human medical ground in practice. This makes me capable of also using my biomedical and diagnostic knowledge with discrimination and judgment.

Nine

TRANSGRESSIVE TECHNOLOGIES IN REPRODUCTIVE MEDICINE: DO THEY CALL FOR A REVISION OF NOTIONS OF HEALTH?

Kristin Zeiler

1. Knowledge that Unveils

Human beings act in and upon the world. Action yields experience, reflection, and knowledge, and knowledge means awareness of possibilities and risks. Today, the perceived limits of human knowledge are revised and reshaped in the field of medical genetics.

Developments in medical genetics such as reproduction technologies push the limits of common understandings of the goals of medicine. Possible goals of medicine are no longer only the cure of diseases experienced in a human being's everyday life (as in classical "bedside" medicine) or prevention of diseases in a born human being; they also include the prediction of diseases in an embryo. We can understand cure, prevention, and prediction as different parts of a general goal of medicine: "promotion of health." The promotion of health is not the only possible goal of medicine, but due to limited space, the article will discuss only this general goal of medicine in any depth.

The article will discuss two *in vitro* fertilization-related technologies, namely pre-implantation genetic diagnosis (PID) and germ-line gene therapy (GLGT). The aim of PID is to identify the presence of genes that will or might result in genetic diseases in an embryo. The technique diminishes the risks of unconsciously implanting an embryo with such genes. "Healthy" embryos can be implanted. The fate of GLGT is yet unknown, but its general aim is to modify genetic material in germ-line cells before implantation. Again, "healthy" embryos can be implanted.

The use of PID and GLGT appears to promote health in some sense. When we analyze the notion of health, we begin to unveil the ambiguity of that statement. The purpose of this article is to discuss consequences of three clusters of contemporary health theories or definitions of health when applied to reproductive medicine, and to explore the limits of promotion of health as a goal of medicine. Reproduction technologies, I will argue, highlight ambiguities in the concept of the "patient" or the promotion of health as a medical goal.

The first cluster of theories describe health as normal, as the biologically normal functions of organs or as the statistically most common state of being. The second cluster involve holistic-idealistic definitions of health, and the last cluster focus on notions of health as capability. I will discuss these notions of health from the perspectives of the potential parents' health and the health of the embryo. I will use excerpts of interviews with geneticists and gynecologists that formed part of the project "Lifeviews, Human Nature, and Genetics." A fictitious example will serve as introduction.

A. Anna and Peter

Anna and Peter arrive late to the clinic. For the third time this month, they park the car close to the main entrance and walk slowly up the alley. The last weeks have lain heavy on them: Their minds are occupied with discussions, questions, guilt, and hopes. Dreams of what could have been—but is not—have been floating around them. As Anna puts it when they sit down with the doctor at the IVF clinic:

> We don't regret what we've done. We wouldn't have acted differently if we could have turned back time. But we have more knowledge of what kind of life a second child probably would have now, and we have a little better feeling of what we can handle, and of what we … can't.

Peter nods and adds,

> We've not been discussing much else these last weeks. How would it feel to have another child with the syndrome? I mean, our first daughter needs extra time from us now, and she is only six years old. What would be best for her, to have a younger sister or brother with or without the syndrome? Do we really want to go through IVF? Well, we do want one more child, but … We believe that two children with fragile-X would be too much, for us and for them. We wouldn't have strength or time enough to give them what we want and what they need.

He hesitates, and the room is silent. Anna picks up the discussion again.

> We would like to try the method that you mentioned and try to have a child without the syndrome. We would like to use PID.

2. The Notion of Health in Reproductive Medicine

A. Health as Normality

"Isn't it possible that mothers possess a right to a small slice of normality or to the normal joys of motherhood?" one interviewed gynecologist asks. She discusses the situation of a single mother with two children with fragile-X, an x-linked genetic disease that results in walking difficulties as well as mental disabilities and difficulties with reading and speaking. The mother is a carrier of fragile-X syndrome. She has recently re-married, and she and her new husband want to have a child together. The woman must be exhausted, reflects the gynecologist. She knows that the first two children will not live autonomous lives. The gynecologist sighs and says that she would understand and support such a mother if she wanted to use PID. Even though it sounds ugly to say so, she remarks, could not a right to normality exist (Interview 4)?

The first cluster of health theories and definitions of health center, in different ways, on the notion of "normal health." In the gynecologist's reflection, a healthy child is normal in some sense. From a theoretical point of view, we might interpret "normal health" as the biologically normal functioning of an organ or biomedical unit or as the statistically most common state of being. In both cases, "normal health" might be descriptive as well as normative. Definitions of normality can be used as a screen through which to discern and sort out the unwanted.

We can understand Christopher Boorse's theory of health as a variant of the first interpretation of normal health. Health is the lack of impairments or limitations on "normal functional ability." An organ is healthy when it functions as it would under "normal circumstances." Health is defined *via negativa* as the absence of disease. Boorse defines a disease as "a type of internal state which is either an impairment of normal functional ability, i.e. a reduction of one or more functional abilities below typical efficiency, or a limitation on functional ability, caused by environmental agents" (Boorse, 1997, pp. 7–8).

Diseases are subnormal functions or dysfunctions of an organ or a system of organs, either on the level of cell organelles or complex human behavior. Normal function is the "statistically typical contribution by [an age group of a sex of one species] to their individual survival and reproduction," and species design "is the internal functional organization typical of species members, which (as regards somatic medicine) forms the subject matter of physiology" (Boorse, 1997, pp. 7–9). One of the major quests in Boorse's writings is to define health and disease value-neutrally. Phrases such as "to be healthy" or "to have a disease" should not be understood as *a priori* statements of wanted or unwanted conditions. They are non-normative descriptions of how a bodily function is performed (Boorse, 1981, p. 548).

A different approach to health would be to define it "in reference to a tacit norm, signified by '*the* human genome', and in contradistinction to a state of unhealth (or abnormality), indicated by an ever-growing list of conditions characterized as 'genetic disease' " (Fox Keller, 1992, p. 297). In this approach, which Ellen Fox Keller characterizes and criticizes as the rhetoric of genetics, health is defined as consistency with a microbiological norm: the norm of DNA.

If disease is a statistically uncommon state of being, what counts as health and disease would fluctuate with changes in the genetic structure of people. Since most human beings have some kind of genetic variation on a structural level, we could argue that it is healthy, that is, statistically normal, to have some kind of genetic variation. We could also define health as correlation with the statistically most common nucleotide sequence that constitutes a gene, but such a position would be difficult to maintain, since nucleotide sequences are characterized by (slight) interpersonal variety, contributing to the variation in human traits observable in a population. (A gene is a string of nucleotides, at a location on a chromosome that encodes a product, a polypeptide, or an RNA product. Some variations of genes may result in genetic diseases, while some will not.)

A more fruitful interpretation of health on a genetic level would be to focus on the concept of a gene as a segment of DNA with an assigned function. We could understand health, then, as absence of dysfunction in the sequencing, replication, and transport related to DNA-sequences encoding for proteins, and we could view that interpretation as a special case of the Boorsian definition. The European Union Commission's 1988 proposal also referred to a genetic concept of disease, describing "predictive medicine" as practice that seeks "to protect individuals from the kinds of illnesses to which they are genetically most vulnerable and, where appropriate, to prevent the transmission of genetic susceptibilities to the next generation" (quoted in Koch, 1999, p. 188).

Jerome C. Wakefield provides my last interpretation of health in terms of biologically normal functioning. A disorder, the term Wakefield uses to cover traumatic injuries as well as wanted or unwanted disease, is best defined as a "harmful dysfunction." Dysfunction in Wakefield's usage is a "factual" term that refers to "the failure of an internal mechanism to perform the function for which it was designed." It is a hindrance to the "natural function" of an organ, which is developed through evolution (Wakefield, 1992, p. 382). Such a description has links to the Boorsian account of disease, but Wakefield adds the value-laden concept of harmfulness: Disorders cause harm to individuals in the environmental and cultural setting in which they live.

Value-neutral definitions of health and disease function as a watershed, with Boorse and Wakefield on either side. In the context of choices in reproductive medicine, the value-neutrality of health becomes especially difficult to grasp. The choice to use PID or GLGT is entrenched in values. Once made, the decision, to use the techniques or not to, is a reaffirmation of the value of health, directly or indirectly, in itself or instrumentally. Health status is what distinguishes embryos from one another. Boorse explains that his concept of disease is that of a pathologist, not a clinician, because in a clinical context, any notion of disease or health becomes value-laden (Boorse, 1997, pp. 6, 17).

i. Normality Applied to Reproduction

I will assume that the goal of medicine is to promote health in the sense of biologically normal functions of an organ or a gene. What consequences would this definition have when applied to reproductive medicine? Can we understand PID and/or GLGT as means to promote the potential parents' health or the embryo's health in this sense?

If Anna and Peter, in the fictitious situation, had been unable to become pregnant for medical reasons, then one or both of them would have had a disease in the sense of biological dysfunction. If medicine promotes health, if health is the lack of disease, and if everyone is entitled to health-care treatments, Anna and Peter could have claimed the right to *in vitro* fertilization, IVF. In Boorse's vocabulary, Anna and/or Peter would have a "therapeutic abnormality" requiring treatment (Boorse, 1997, pp. 6, 17).

In the case of PID, couples need not be infertile to use the technology, and they need not have a disease in the sense of biological dysfunction. This is Anna and Peter's situation. Their organs function as they should under "normal circumstances," and they are healthy in the biologically functional sense. Reproductive technologies cannot be understood as means to promote their health. One exception exists. If being a carrier counts as having a disease, then Anna and/or Peter could be regarded as patient(s). Such an expansion of the concept of disease would evoke questions of how to define a carrier. A carrier has genes for conditions that will not be expressed in the phenotype—she or he does not have experienced biological dysfunctions. Within the framework of health as biological-functional normality, we cannot find reasons for using PID in order to treat potential parents.

Can we understand PID as a way to promote the embryo's health? If so, the potential embryo might be seen as a potential patient. Such a view needs to be clarified. When Anna and Peter decide to enter the clinical program of PID, Anna's eggs are not fertilized. No embryo exists, although we might hold that potential embryos exist. Two aspects of PID complicate viewing the potential embryo's status as a potential patient as a reason for treatment. First, once fertilized, Anna and Peter's embryos are at risk from fragile-X, but they will not necessarily develop the disease. They may not even become carriers (that is, if both parents are not carriers). Consequently, if embryos are potential patients, they would be so due to the possibility that they will develop a disease, not because they necessarily will do so. This double potentiality—a potential embryo with a potential disease (who might be regarded as a potential patient) —complicates the matter of agency.

Second, if we see PID as a method for the promotion of health, then we must understand and accept the destruction of some embryos during screening as part of the "treatment." This leads to another question of agency. Embryos chosen for screening—that is, treatment—are consequentially destroyed. Embryos chosen for diagnosis are not the ones that will be implanted, and that might be carried to term nine months later. If embryos are potential patients, we need to distinguish between embryos screened during PID and genetically identical embryos that are not screened, but are implanted.

The second technology discussed in this project, GLGT, would fit better into the discussion of an embryo having a disease in the sense of biological dysfunction. Doctors would perform GLGT on an existing embryo with known genes that result in diseases.

If the goal of medicine is to promote health, an agent must exist whose health the treatment promotes. From the view of health as biologically normal functioning, Anna and Peter have no disease. Within this framework, Anna and Peter would not qualify as patients in need of health care. The question of a potential embryo with a potential disease leads to ambiguities in how to understand agency—a necessary step in resolving whether the potential embryo should be considered as a patient.

Physicians can hardly perform PID for the sake of promoting health in the biological functioning sense. If PID is to be part of the health-care sector, we must abandon the use of health as biologically normal functioning, or, as is sometimes done, argue that health care should not only promote health in this sense, but also promote welfare or happiness. The only situations where disease in the sense of biologically abnormal functioning is present is in occasions where a couple uses IVF as a necessary means to becoming pregnant or when an embryo exists with a genetic disease; that is, in cases of IVF or GLGT.

What if we understand the goal of medicine as promotion of health in the second sense described here, as the statistically normal state of being? Such an understanding of health is likewise difficult to maintain when applied to reproductive technologies. At present, the birth of children without technical assistance is statistically normal, healthy by definition. Given the understanding of health as the statistically most common state of being, we could argue that most of us are born with some kind of genetic variation, and that variation—even variation that results in disease—is (per definition) healthy. Following such a line of argument, little reason exists for preventing genetic variation from the perspective of health-promotion. Furthermore, new technologies or technologies that apply only to a minority will always be uncommon and statistically abnormal.

Notions of health as the statistically most common state of being or as biologically normal functioning can be used in reproduction technologies, but not without difficulties. Reproduction technologies uproot the question of agency and the doctor-patient relationship in the classical medical setting. If PID and GLGT are parts of the health-care system, a broader notion of health than those discussed here is needed, or medicine needs to count the promotion of welfare and happiness among its goals—aside from the goal of promoting health in any sense of "normality." Therefore, I will move on to a second cluster of health theories.

B. Sanity, Sanctity, and Salvation

The second cluster of theories and definitions of health are both holistic and idealistic: They are holistic in the sense that they emphasize not only biological, but also psychological and spiritual dimensions of health, and they are idealistic in the sense that they define health in a manner that makes it more or less unobtainable during a human life. The World Health Organization definition is the example *par excellence* within this cluster. It sees health as "a state of complete physical, mental and social well-being and not merely the absence of disease and infirmity" (*Official Records of the World Health Organization*, 1946). To the extent that similar theories are idealistic, they are also value-laden just as ideals are value-laden. The definitions are both descriptive and normative, describing and proclaiming desired states of being.

Theologians in the 1960s and 70s elaborated the two other examples of holistic-idealistic approaches to health discussed here. The emphasis on holistic approaches in this sense, and on "spiritual issues" in health care, re-emerged in the 1990s in British discussions of nursing in hospices (see, for example, Robert Lloyd-Richards, 2000)

Bernard Häring, a Roman Catholic theologian, bases his definition of health on anthropology. In Häring's view, the human vocation is to affirm the "total" human nature: the human being-in-the-body, being-embodied spirit, and a body filled with spirit, open "to the Other" in human beings (Häring, 1972, pp. 50–51). Human health, or "redeemed health," as Häring phrases it, entails "the greatest possible harmony of all man's forces and energies, the greatest possible spiritualization of man's bodily aspect and the finest embodiment of the spiritual." The ideal dimension is obvious. Health appears to be a kind of integration between different aspects or dimensions within human beings. Health involves a wide range of the dimensions of a human life, not only biological ones.

A similar emphasis, but from an existentialistic point of view, is present in Paul Tillich's description of health. Tillich does not start from an idea of human nature in Häring's sense, but states that just as human life is a "multidimensional unity," so is health. Tillich argues that we can achieve health in different dimensions, and that we might be more or less healthy in these different dimensions. In Tillich's view, someone might obtain health in a "mechanical dimension" through surgery, and health in a "psychological dimension" through a process of promoted self-awareness, self-alteration, and enhanced integrity. A person might approach health in a "spiritual dimension" when the human spirit transcends to or through the Divine and the Holy. We must attain all aspects of health—and those aspects mentioned here are just some of the aspects that Tillich describes—to achieve health in its fullness, as "complete health" (Tillich, 1967).

The WHO definition, as well as Häring's and Tillich's approaches to health, allows the human experience of bodily disease to co-exist with feelings of health. The definition emphasizes the importance of different dimensions of health, or of layers in the experience of being healthy. In comparison to the WHO definition, Tillich's and Häring's notions are graded.

Health in holistic-idealistic approaches is overtly value-laden. Health transforms into an ideal when we describe it as a part of human nature that aims at true humanhood. In Tillich's view, health is similarly value-laden. Complete health is obtained only in a "fulfilled" human existence. These definitions also contradict some forms of pure materialism. To be healthy is not only a question of biological functions. Tillich's and Häring's approaches to health could be described as ways of living; they are processual. All theories or descriptions of health are more or less eschatological. They are left unfulfilled during a human life span.

i. Holism and Health in Medical Praxis

What consequences do holistic approaches to health have when applied to reproductive medicine? I will, once more, assume that promotion of health is a key goal of medicine.

If health, as defined by the WHO, in Häring's "redeemed health," or in Tillich's "multidimensional" sense, should be promoted, then the goal of medicine would need to be broad. This call for broadness is a major contribution of holistic theories to discussions of the goals of medicine. To understand the complex nature of health, we need several categories and perspectives. Health is not only biological-functional normality. We must discuss existential, psychological, social, and/or religious dimensions of a human life, since we, according to these definitions, might be healthy or unhealthy in any of these areas.

What practical consequences would holistic definitions of health have when applied to PID and GLGT? Would Anna's and Peter's health, or the potential embryo's, be promoted through PID or GLGT that resulted in a healthy daughter?

In Häring's words, Peter and Anna would have redeemed health if they lived in accordance with their "total human vocation," if body and spirit were integrated parts of their lives. This sketchy description of the human being has no obvious implications for choices in reproduction technologies, even though Häring indicates that we should not consider conscious interventions in the human genome immoral as long as the "whole man" is at the heart of the ethics. Humans are free to shape themselves actively, even through such means as genetic engineering. What Häring has in mind appears to be GLGT, not PID; that is, a procedure with no loss of embryos (Häring, 1972, p. 64). At the heart of Häring's view lies a concern with the direction of a human life—the directedness to the human *telos*—not, primarily, a concern with the designated technologies.

The use of Tillich's notion of health in medical praxis gives rise to a similar discussion. Anna and Peter would have "complete health" if they were healthy in all dimensions of their lives. Such completeness appears too idealistic to be a goal of medicine. We could argue that Anna and Peter do not have biological health if they cannot have children without medical assistance. If that is so, then we find ourselves back at the previous discussion of biological normality. But this is not Anna and Peter's case; they are biologically healthy. We might also argue that the birth of a second child might promote Anna's and Peter's psychological health. They might experience less strain on the family as a whole, but that is not a necessary consequence. A second child with fragile-X is no *a priori* argument for the way that Anna and Peter will experience their life situation with such a child, and it could hardly be a universal argument for parents' rights to reproductive technologies.

The argument of promoting psychological health through the use of reproductive technologies runs into other kinds of difficulties as well. Anna and Peter might argue that they are not completely healthy, in a psychological sense, without a healthy child or a child with certain characteristics. The third cluster of health theories, to be discussed below, will bring up a similar line of argument, which will lead to such questions as, to dicussed bellow, "What kinds of requests are 'reasonable' for the health-care system to deal with?"

The use of PID or GLGT will not necessarily enhance Anna's and Peter's health in the holistic sense. Could we argue that PID or GLGT would promote the embryo's or the potential child's health? A first problematic issue is the question of how we can consider the embryo's health as promoted through termination in PID. In GLGT this would not be a problem, since the aim is to modify nucleotide sequences within existing embryos that will be implanted in the uterus.

A second problematic issue is just as relevant to holistic approaches to health. Being born with or without a disease does not settle the question of whether we are healthy, except in the WHO definition (where health requires complete well-being). In Häring's and Tillich's views, health can be a graded concept: We might still be healthy with a biological disease. Lynn, Anna and

Peter's first daughter, might not have complete health in Tillich's view, but she might be healthy to some degree, or in some dimensions. From the holistic point of view, whether Lynn is healthy does not depend on her medical diagnosis.

The broadness of holistic health definitions is one of their pitfalls, as expressed in the frequent criticism of the WHO definition. Health is a desirable but unreachable ideal. A similar criticism could be directed at Häring's and Tillich's approaches. Most human experiences would be gathered under the label of health, in Häring's view, or the notion of "complete health," for Tillich. In a theological context, to be healthy and to be human appear to be synonymous with being "saved." This is explicitly so in Tillich's writings, where salvation is understood as both temporal and eternal, an ongoing process. "To announce that salvation is at hand and to heal is one and the same act," argues Tillich (Tillich, 1984, pp. 19–20). We may criticize the holistic approaches for being falsely broad—especially Häring's notion of "redeemed" health, which is built on a theological anthropology and presupposes a trialistic view of human nature.

Holistic interpretations of health run into difficulties when it comes to concrete choices regarding the use of both PID and GLGT. The definitions of health are vague, and so will be any discussions of the consequences of an action. A reflection by one of the interviewees presents an exception. According to this interviewee, a "true" value-hierarchy exists, as well as a true human nature that human beings must keep to. Doctors should not perform PID since it presupposes a false idea of what should be characteristic for human beings (life as a gift), and of when a person is human (the discussion of the ontological status of an embryo). This argument focuses on which aspects of a human life might be lost if we use screening, which ideals are promoted, and which are not (Interview 2). A holistic definition of health has practical implications when related to this view of human nature.

The purpose of the theories mentioned here, we might argue, is not to provide guidelines for the circumstances in which a couple is in need of health care. Such questions must be treated from case to case; then, we can use holistic notions of health in the same context as virtue ethics and contextual approaches to ethics. The idea is not to establish applicable principles or guidelines, but, among other things, to widen the discussion of what health is.

A crucial question in holistic accounts of health such as the one put forward by Häring is what it means to be an integrated, or "whole," human being. Would "to be healthy" be "to know oneself"? Or, in the vocabulary of Charles Taylor, would healthiness be helping people find their orientation in the moral space, "a space in which questions arise about what is good or bad, what is worth doing and what is not, what has meaning and importance for you and what is trivial and secondary" (Taylor, 1989, p. 28)? If so, holistic theories would have definite implications for genetic counseling.

Theorists sometimes criticize medicine for not being holistic enough. Difficulties with "holistic medicine" might be due to the vagueness of holistic approaches and the difficulties in applying holism to the outcome-oriented structure of health-care systems. Holistic approaches to health are useful in discussions of reproduction technologies in the sense that they broaden the discussion. Notions of health, however, provide no obvious guidance in what to do in reproductive medicine.

C. Health: A Question of Capability

According to one of the interviewees, to be healthy is to be able to interact, to discuss, and to joke, while someone else would likely define health differently (Interview 4). Reflections like this one conceive of health as a personal concept, colored with meaning that depends on personal experiences and convictions. Such a view can elaborate no objective or substantial definition of health. Another interviewee uses the notion of well-being instead of health. In her view, to be well is to be free, whether in terms of possibilities to act, to think, or to choose freely (Interview 3). Both interviewees are skeptical about the possibility of defining health in substantial terms. However, their reflections on the notion of health appear to focus on capability to act: the ability to interact, the ability to choose freely. In the last cluster of health theories, health is the individual's ability to perform or obtain things held to be valuable. We could label such views as holistic, like the second cluster of health theories, in the sense that health is not only an attribute for a designated organ, but a concern of the whole individual. The following will describe two theories that would fit the description "holistic capability."

According to Caroline Whitbeck, health is "a person's psychophysical capacity to act and respond appropriately (in a way that is supportive of the person's goals, projects and aspirations) in a wide variety of situations." To be healthy is to be capable of acting and reacting in a manner that supports the goals that a person might have. Whitbeck argues that the notion of health is "strongly value-laden." To use the notion of health, she argues, is "to say that that to which [health is] applied is good or bad, valuable or worthless, in some way" (Whitbeck, 1981, p. 614). In Whitbeck's view, the notion of health is indirectly related to happiness. In order to be happy, a person "needs to be able to act in ways that serve many goals, aspirations, and projects simultaneously." The possibility of being happy is seen as a function of, among other things, health (Whitbeck, 1981, p. 620).

Lennart Nordenfelt elaborates a second example of capability approaches to health. In his approach, health is a "person's ability, in standard circumstances, to reach his vital goals" (Nordenfelt, 1987, p. 126). Again, to be healthy is not to reach crucial goals, but to have an ability to obtain what is "vital." The notion of health is ability-centered. Some notions used in this definition need clarification. In Nordenfelt's view, the substantial content of vital goals may differ with individuals, even though some basic needs are universal. Vital goals are "those state of affairs the realization of which are necessary and jointly sufficient for [the individual's] minimal welfare" (Nordenfelt, 1987, p. 126). The question of exactly which goals qualify as vital are entangled with epistemic difficulties. We do not have "complete epistemic contact with [our] emotional life" (Nordenfelt, 2001, p. 105). Our conscious judgment of what is a vital goal only captures what we can articulate in verbal terms, and we might hold unconscious vital goals.

In the human case, Nordenfelt equates minimal welfare with minimal long-term happiness. Minimal long-term happiness springs from the fulfillment of vital goals; it is the result of the realization of vital goals. To put it differently, minimal happiness is the equilibrium between realized vital goals and reality as the individual experiences it. Nordenfelt emphasizes that it is possible to be healthy and unhappy simultaneously. The crucial point is to be capable of happiness, to have the capability to realize vital goals. Just as the capability to reach vital goals might be stronger or smaller, health is a graded concept.

A third notion Nordenfelt discusses is "standard circumstances." Standard circumstances differ with nations and cultures. They are those circumstances under which you might expect to live in a given culture. In his later writing, Nordenfelt adds the notion of "accepted circumstances" which may designate either standard circumstances or such circumstances as the individual him or herself considers as reasonable (see Nordenfelt, 2000, pp. 70–73; Nordenfelt, 2001, p.68).

Similarities exist between these theories. Both define health independent of bodily disease, as an instrumental value necessary for the acquisition of the described goals. Nordenfelt's theory of health is indirectly value-laden, in the sense that it is a capability to reach a goal necessary for minimal happiness. As indicated above, Whitbeck argues that her notion of health is "strongly value-laden." Furthermore, both forms of capability theories presuppose some degree of freedom. If health is the capability to obtain vital goals, or the capability "to act and respond appropriately," the individual needs to have some level of external as well as internal freedom as part of the capability. Finally, both theories presuppose bonds between capability of health and happiness. In Nordenfelt's view, realizing vital goals is sufficient for obtaining minimal welfare—that is, in the human case, happiness. For Whitbeck, health is one among many causes of happiness.

i. Healthy Children and Health as Capability

I will, once more, assume that the goal of health care is to promote health. The last cluster of health theories emphasize that health is not only the absence of biological (or psychological) disease. In Nordenfelt's theory, medicine enhances health as long as the individual's capability to fulfil vital goals is strengthened. What would this understanding of the notion of health mean when applied to the use of PID or GLGT?

Anna and Peter long for a healthy second child. They love their first daughter, but they want to enlarge the family with another biological child. Anna and Peter are well aware of the chance that a second daughter might be born with fragile-X, and they are strongly inclined to use PID in order to secure the birth of a healthy child. They know that they cannot be sure that they will become pregnant through IVF, but they feel that they want to try.

According to Nordenfelt's theory, vital goals are events or conditions that, when realized, result in minimal long-term happiness. If a healthy second child is not necessary for Anna and Peter's minimal happiness, then a second child is not one of their vital goals. If health-care systems should promote health as the capability to obtain all vital goals, then either the birth of a healthy second child would be on the medical agenda (and would count as a vital goal), or the health-care system should work for promotion of things other than vital goals. Similar reasoning would apply to Whitbeck's notion of health.

Let us assume that a healthy biological child is necessary for Anna and Peter's minimal happiness; a healthy child is one of their vital goals. Anna and Peter are not completely healthy in the capability sense, even though they have no disease or biological hindrance to getting pregnant. Let us assume that they enter the IVF and PID program and that an embryo is successfully implanted and carried to term. They have been given the capability to obtain their vital goals through technological means. They are healthy.

We might include other examples of reproduction technologies in this discussion. Envision a future scenario: Anna and Peter enter the IVF clinic. They describe their situation and the doctor reassures them that they have come to the right place. GLGT is performed at the clinic under rigorous medical surveillance. If Anna and Peter want to, it would be possible to modify nucleotide sequences once the egg has been fertilized, but before it is implanted in Anna's uterus, so that a healthy child will be born.

Let us assume, for the sake of testing the consequences of capability theories, that Peter wants an intelligent daughter with brown eyes. Would such a request still fall into the realm of vital goals? The doctor might try to find out how central such a goal would be. Is it just a momentary desire? Is it really necessary and sufficient for Peter's minimal happiness? Would Peter have a reduced health in the capability sense if he did not have an intelligent, brown-eyed daughter? Peter and Anna have blue eyes, and due to their genome they would not have the capability to fulfill Peter's vital goal without medical assistance. It appears that little restriction can be put on uses of technology that medicine should develop.

At this point, we return to what Nordenfelt labeled as accepted or standard circumstances. Would PID and GLGT be standard circumstances in the future? Reproduction technologies might not yet be called standard circumstances, but that is a matter of how common (or uncommon) they are. If a healthy, intelligent, brown-eyed daughter were a vital goal in the sense elaborated by Nordenfelt; if the goal of medicine is to promote complete health in the capability sense; and if these reproduction technologies would qualify as standard circumstances, then no reasons against allowing GLGT for human characteristics exist within the health theory.

We might criticize capability theories for having overly strong bonds between happiness and health. Nordenfelt assumes that everybody has "unconsciously held vital goals," which, when realized, result in minimal happiness (Nordenfelt, 2001, p. 108). But is fullfilling all vital goals sufficient for obtaining minimal happiness? In Nordenfelt's view, this is, by definition, the case. But we might ask if it is reasonable to link health to happiness in this way. As posed by Nordenfelt himself, does "the healthy person have all internal resources needed for his or her state of minimal happiness?" (Nordenfelt, 2001, p. 109). Happiness, we could argue, is not only equilibrium between realized vital goals and perceived reality, as in Nordenfelt's theory. We may also criticize capability theories for their consequences in the example above. Within this framework, no dividing line can be drawn between treatment of diseases and enhancement of an embryo's characteristics.

So far, most of this discussion of health has focused on the woman and man who want to use a reproduction technology. A discussion of the capability to realize vital goals is difficult with respect to a potential embryo. Potential embryos can hardly be considered as having capability to actions; they possess only potential capability. From this framework, any decision to use PID or GLGT must be paternalistic. We do not know what the vital goals of the potential embryos are. The alternative is to argue that we should perform PID and GLGT as a step in promoting the potential parents' health.

Capability theories of health have implications for reproductive technologies. They allow an experience of health that can co-exist with biological disease. If a goal of medicine is to promote complete health in a capability sense, then the health-care sector should provide PID and GLGT, as long as healthy children are vital goals. We cannot draw restrictions on technologies from the goals set up for health-care systems, given these health theories.

3. Conclusions

The notion of health is crucial to any discussion of the goals of medicine. When applied to reproductive medicine, different notions of health sometimes have contradictory consequences. The definitions we use color our discussion of whether we can see PID and GLGT as means to promote the health of potential parents or embryos.

Promotion of health is not the only possible goal of medicine. In a general sense that includes cure, prevention and prediction, we might understand the promotion of health as a candidate for the label "goal of medicine." Reproduction technologies highlight the limits of some understandings of this possible goal of medicine and call for either a broad notion of health or for other goals of medicine. Reproductive technologies also call into question our understanding of the notion of the patient.

If the goal of medicine is to promote health in any "normal health" approach (as in the first cluster of health theories), reproductive technologies would not be part of the medical agenda for couples in Anna and Peter's situation. Such technologies would not promote health—as the biologically normal functioning of an organ or as the statistically most common state of being—in such parents as Anna and Peter. They do not have a disease. Furthermore, promotion of health for implanted embryos brings up difficult questions of agency. This is not the case in GLGT, performed on the embryo on its own behalf, but GLGT is strictly forbidden in most European countries. In Sweden, not only is GLGT forbidden, but also research that might lead to the development of technologies for GLGT (*Lag om åtgärder i forsknings- eller behandlingssyfte med befruktade ägg från människa*, SFS 1991: 115). In the second sense of health as the statistically most common state of being, it would be likewise difficult to argue that the use of PID or GLGT would promote health.

To endorse any of the "holistic-idealistic" approaches to health, reproductive medicine, as well as medicine generally, needs to focus on the psychological, existential, and religious dimensions of a human life, as well as on biological ones. We might perform PID and GLGT, but it is difficult to argue that they necessarily promote the health of the potential parents or the embryo implanted. Holistic definitions of health do not give much guidance in concrete life situations.

Finally, to support any of the "holistic capability" approaches, it would be the task of medicine to perform PID and GLGT for the sake of the potential parents. Within this framework, we could hardly argue that we should allow the sorting out of embryos carrying genetic structures for some diseases, but not for others. Enhancement of intelligence and determining eye color might be part of "vital goals." If health is the capability to fulfill vital goals, it would be hard to draw a line between diseases and other characteristics.

The promotion of health as a goal of medicine is problematic when applied to reproduction technologies such as PID and GLGT. This might be due to the different notions of health involved, or to the technologies themselves. PID and GLGT could be understood as "transgressive technologies" in the sense that they question and transgress the present understanding of the limits and goals of health-care systems, as well as the common understanding of the concept of the patient.

Interview excerpts come from interviews performed in Italy May, 2000. All interviewees are anonymous. Supporters of this study are The Swedish

Foundation for International Cooperation in Research and Higher Education (STINT), The Swedish Foundation for Strategic Research (ELSA), and The Church of Sweden, Department of Pastoral Development.

Ten

OLD, ILL, AND ESTRANGED: ON THE NEEDS OF PERSONS RECEIVING ELDERLY CARE

Bengt Richt

For many years, much of the work in the field of elderly care has dealt with meeting the needs of the elderly. In Sweden, Abraham Maslow's half-century-old model of needs (Maslow, 1954) has come to function largely as a guide for action in this area of care (Falk, Gustavsson, and Richt, 2001). The advantage of such a framework is that it expands the focus of care from a narrow concentration on the biological organism to the individual in his/her totality. The disadvantage is that the model is too general to provide guidance in an individual case. In practice, Maslow's model has therefore either come to lend itself to different ideological ends regarding care or degenerated into the idea that the duty to provide the elderly with personal service belongs to health-care staff.

This responsibility has often been synonymous with carrying out different practical chores, such as cleaning and keeping things tidy, doing bank errands, or helping the elderly make purchases. It is obvious—or it should be—that the old person is offered help with personal tasks that he or she cannot do independently. This kind of personal service, however, covers only a part of the person's total needs. The current focus of elderly care is not only disadvantageous for those receiving care, it also puts the staff in a position that is dangerously close to the humiliating role of a maid, and risks affecting the prestige and attractiveness of their work.

The large majority of those receiving formal elderly care are also ill. Illness shifts and changes the needs that characterize the totally healthy state. One reason Maslow's model of needs—as well as other influential and general theories, such as Erik Homburger Erikson's theory of development (Erikson, 1982)—are inadequate with respect to providing guidance in practical care is that they do not take into consideration what it means to be ill. Thanks to developments in recent years in health philosophy and medical sociology, however, we know enough today about the phenomenology and social psychology of illness to be able to fill the otherwise rather empty framework of needs with something more concrete. In this article I will review some of the most widespread conceptual frameworks in the field, after which I will draw some implications for work in elderly care. Because my background is mainly in medical sociology, I have found it natural to refer first to this area. Still, rational considerations also exist behind my choice:

(1) Generic concepts have developed, primarily within medical sociology, regarding the understanding of illness. The philosophical-phenomenological treatment of the subject has often had the character of applied phenomenology.
(2) Purely phenomenological analyses have often focused on separating the phenomenological perspective from a biomedical perspective, a little to the detriment of the true analysis of illness. Medical sociology has largely outgrown this controversy with medicine.
(3) Medical sociological conceptualizations are often based on phenomenological, interactionist, and sometimes narrative grounds, and thereby include some philosophical insights. The reverse is seldom the case.
(4) Medical sociological conceptualizations (even if they are often less lucid) are focused more on variations and processes than their philosophical counterparts. Therefore, philosophy has something to learn from sociology and something to convey to sociological concept- and theory-development (for a positive model, see Nordenfelt, 1995a).
(5) In its more complete separation from traditional medicine, medical sociology has unnecessarily neglected to draw the practical implications of its insights. The need for explication is therefore greatest in this area.

Before turning to the main objective, I will describe the thinking that has characterized elderly care since the 1960s to justify further the need for an amplification of the currently dominant frameworks of thought.

1. Perceptions About Good Aging

One theory—or rather, ideology—that has long permeated care is the so-called activity theory. The theory is based on the premise that the individual's quality of life is positively correlated with his/her degree of activity. "The greater the physical and social activity, the greater the life satisfaction" is the motto for care based on the reasoning of the activity theory. Younger or middle-aged healthy, occupationally active individuals naturally perceive of the activity theory as intuitively reasonable. Who wants to live a passive life in withdrawal and social isolation? In addition, this perception is underlined by the finding that a rich social and physical life has a positive impact on a person's state of health.

But other ways of looking at aging have challenged the activity theory. During the 1960s, the disengagement theory (see Cumming and Henry, 1961) had an upswing. According to this theory, inactivity in old age is normal and associated with a high level of life satisfaction. The theory is based on three hypotheses. The first hypothesis is that in one way or another, society pushes old people to the sidelines. The second hypothesis is that the individual screens himself or herself from society. This disengagement has social and mental dimensions. Social disengagement refers to decreased interaction between the elderly person and his or her previous social network, while mental disengagement refers to the elderly person's decreased interest in people and events in his or her surroundings. The third hypothesis is that despite a decreased number of roles, the individual can experience a high degree of life satisfaction.

The gerotranscendence theory (see Tornstam, 1997) offers still another way of looking at aging. According to this theory, aging does not necessarily have to be a continuation of an earlier lifestyle, but can instead involve development toward something new and different. The theory is based on the idea that the aging individual often exchanges his/her earlier, rational, activity-focused attitude for other values and another way of being. Activity in its traditional sense would not be the universal path to good aging. An early forerunner of the theory of gerotranscendence is the psychologist Carl-Gustav Jung, who in the essay *The Stages of Life* expressed the following about the problems of aging:

> A human being would not grow to be seventy or eighty years old if this longevity had no meaning for the species to which he belongs. The afternoon of human life must also have a significance of its own and cannot be just a pitiful appendage to life's morning. The significance of the morning undoubtedly lies in the development of the individual, our entrenchment in the outer world, the propagation of our kind and the care of our children. This is the obvious purpose of nature. But when this purpose has been attained—and even more than attained—shall the earning of money, the extension of conquests and the expansion of life go steadily on beyond the bounds of all reason and sense? ... Where is the wisdom of our old people—where are their precious secrets and visions? For the most part our old people try to compete with the young. In the United States it is almost an ideal for the father to be the brother of his sons, and for the mother if possible to be the younger sister of her daughter (Jung, 1984).

What the alternative should be is an open question for Jung. Cultural creativity, preservation of traditions, idealistic activities, self-fulfillment, realization of previously suppressed life goals, self insight, existential searching, contemplation, development of emotional life, practice of religion, and, finally, work aimed at "discovering death as a goal to strive after" are some of the candidates that are discernible in his text. But they cannot be ranked in order of preference, nor should they be considered as comprising a complete list.

What is most significant is not what the alternative is, but that the individual finds—and gets help in finding—a direction to life that is in line with his or her physical, mental, and cultural prerequisites. "I have observed," says Jung, "that a directed life is in general better, richer and healthier than an aimless one, and that it is better to go in forwards with the stream of time than backwards against it."

In the activity theory, aging is considered a direct continuation of earlier life. The disengagement theory is an attempt to explain why some elderly individuals deviate from that pattern. Jung's formulation is critical to civilization. The norms prevailing in Western society lead the aging individual to abandon his or her true needs. What the theories have in common, however, is that they presume that all people entering old age have the same needs and problems. But scientific studies indicate that old age is just as full of nuances as earlier stages of life.

While all these theories are fairly well-known within elderly care, the activity theory has been dominant. Much of the work during the past 10-15 years has dealt with activation, often collective, in the form of outings, bingo, dancing, or gymnastics. Care staff and relatives arrange these activities with the best of intentions. If everyone does not participate in bingo or in going for a walk, this is felt to be something of a failure. But the needs of someone in the final phase of life can be totally different than those extrapolated from (younger) staff members' preferences; nor do they always coincide with what has characterized the elderly person's earlier life.

An elderly man I know had been an enthusiastic amateur gardener all his life. He was offered a room at a nursing home located next to a common garden area. Everyone thought this was the right place for him. But to everyone's surprise, the man declined the offer. The lesson here is that planning a person's care on the basis of a projection of his/her earlier life can be quite destructive. People's physical and mental conditions undergo continual change, resulting in changes in their preferences and interests. This is especially true for someone who contracts a long-term, and possibly serious, illness. The man in the above example had had a stroke and was partially paralyzed. He did not want to live next to the common garden area since this would involve a continual confrontation with his incapability.

2. Becoming Ill and Being Ill

What does contracting a chronic disease mean for the afflicted individual? Michael Bury formulated the sociological conceptualization that is probably the most influential in this field in the early 1980s. Bury (1982) considers becoming ill as an event that threatens the life of the afflicted person in many respects:

(1) Earlier habits and conceptions that were previously taken for granted lose their stamp of being a matter of course;
(2) The event threatens the person's identity and perception of self;
(3) The basis for social interaction between the afflicted person and those in his/her social network changes. A risk consequently exists that some old relationships will be lost. The character of those that remain will change in a more-or-less radical way.

The perception of illness as a biographical disruption focuses primarily on the changes that occur when a person becomes ill. Thus it can be understood as having the same basis as Talcott Parsons's sick role theory and the measurements of life quality that have been so popular in recent years. In both cases, the ill person is considered to be a primarily passive victim of external circumstances. But already apparent in Bury's original article is that he is mainly interested in the action aspects of illness. That interpretation also receives strong support in a later review article, in which Bury warns that a risk exists that the plain enumeration of illness-related difficulties that frequently occurs in empirical studies gives an incomplete picture of the illness process. The difficulties and losses are incontestable, but are not something with which the sufferer lives passively. They can instead be thought of as challenges that, with the aim of normalization, he or she and those close to him or her continually deal with (Bury, 1991). The biographical disruption consequently comprises a series of "losses" and "work" (also see Corbin and Strauss, 1988). In this respect, Bury differentiates three concepts that are often confused with one another: coping, strategy, and style. This distinction is further developed in one chapter of *Health and Illness in a Changing Society* (Bury, 1997).

(1) According to Bury's understanding, coping has to do with the methods the ill person uses to manage his or her situation emotionally and to restore positive self-esteem. It can comprise attempts to deny the illness or to seek consolation by means of comparisons with others whose situation is worse. Coping has a strong cognitive element.
(2) Strategies are the way in which the ill person, often in interaction with his or her family, tries through his/her actions to deal with illness, daily life, and social interaction. An example is withholding knowledge about the illness from others, which can be done to avoid stigmatization.
(3) Style reflects the notion that different people can have different attitudes toward illness. For example, some people more-or-less consistently refrain from properly carrying out their medical treatment and try to live as they did before; others sometimes take care of themselves, and sometimes do not; while for some, the illness can function as a platform for a new lifestyle.

Expressed in philosopher Fredrik Svenaeus's phenomenological language, Bury's version of lived illness is, consequently, not only "unhomelikeness" but "unhomelikeness wanting to get home" (Svenaeus, 1999).

The conceptualization of illness as a biographical disruption has guided many empirical studies on chronic illness (for an example, see Kelly, 1992). It has also acted as a forerunner to the increased interest in recent times in stories of illness and the functions of narratives in connection with serious long-term illness (Kleinman, 1988; Frank, 1995). Although influential in these two diverse ways, the conceptualization is primary a "crisis model," based largely on the course of events around the time the person becomes ill. As Ann-Christine Gullacksen (1998) points out, however, chronic illness is a continuously ongoing event, not a temporary crisis.

In the early 1980s, nursing researcher and sociologist Kathy Charmaz published an often-cited article dealing more patently with the long-term implications of chronic illness (Charmaz, 1983). She asserts that "loss of self" constitutes a fundamental form of suffering for the chronically ill. She states that a limited life, increasing social isolation, being discredited, and becoming a burden are the most salient sources of suffering for the ill person, and also describes how these factors combine with one another, resulting in a vicious circle where valuable parts of the self are lost. Charmaz's main observations:

(1) Many persons with long-term illnesses become increasingly confined to their homes. They therefore live a limited life where they lack the possibility of developing their identity and self-esteem. The illness becomes a focus in life, and values such as independence and individualism become difficult to exercise. Since many chronic illnesses are unpredictable, anxiety also arises concerning what can happen outside the safety of home. Commonly, a sick individual will live a more passive and limited existence than would seem necessary from an objective point of view.

(2) Increasing social isolation follows as a consequence of the limited life. Many previous contacts are lost because the ill person cannot take part in several social activities. If the sufferer is not retired, he or she can also be forced to change jobs or totally stop working. Medical treatment can also easily lead to isolation. It takes time and is an activity that is not shared with others. In addition, since a mutual basis of experience is lacking, the treatment is not something the ill person can talk about with just anyone. Isolation easily causes the person's attention to focus on his or her failing body. In and of itself—but often along with feelings of shame and inferiority—this introversion can easily lead to a negative spiral of further isolation. Since self-esteem is built up and maintained through social interaction, the process will almost inevitably result in more negative self-esteem.

(3) In relationships with others, the ill person will commonly be treated in a discrediting way. For example, many chronically ill persons find that those around them talk about them rather than with them. When relatives or others close to the sufferer are present, people often turn to these individuals rather than the ill person. In addition, the ill person is frequently considered incapable of doing many things without being asked to do so. The significance of such incidents depends on who is doing the discrediting, the situation in which it takes place, and how often the discrediting episodes occur. The episodes form part of the ill person's sense of self if they occur in situations in which he/she feels vulnerable. They are also more devastating if the ill person identifies with those doing the discrediting.

(4) The increasing economic, physical, and mental dependency that often accompanies serious chronic illness causes the sufferer to experience himself/herself as a burden to others. In Western cultures, where values such as independence, individuality, and the ability to take action are of central importance, the illness often results in a powerful blow to the self-esteem. When even simple everyday tasks become nearly insurmountable problems, the experience of being a burden leads to an increased vulnerability that results in conflicts or in a tendency to withdraw from contacts with the outside world.

The conceptions of illness as "a biographical disruption" and as "loss of self" are central to the sociological understanding of the illness state and have points in common with some phenomenological analyses of illness (for example, see Toombs, 1992; and Svenaeus, 1999). They also complement one another. The first focuses on the time when the disease first occurs, while the second deals in a more apparent way with long-term problems. The first is fairly action-oriented, but the second emphasizes different sources of suffering in chronic illness. Worth noting is that Charmaz, in a later elaboration of her theoretical structure, differentiates three modes of the experience of illness: The illness state can be subjectively understood as an interruption in everyday life; it can be understood as an intrusion; and it can immerse the sick individual in a separate world of illness (Charmaz, 1991).

The wider implication of such a formulation is that the conceptualization of illness as "loss of self" cannot be understood as a declaration of an empirical adherence to law. The condition in which the sufferer is immersed in his or her illness is probably the only one in which loss and suffering are total. Keeping in mind that many chronic illnesses are serious right from the beginning and that others are of a progressive character, "loss of self" is a tendency. The same is naturally the case for the biographical disruption.

3. Courses and Trajectories of Illness

A means of handling disease-related variations and avoiding the overgeneralization of conceptualizations is to talk about different courses of illness. For example, diseases can be stable, on a downhill trend, on an uphill trend, or fluctuating back and forth. The course of illness can be discussed from a disease perspective or an illness perspective. In the disease perspective, the course of illness is described in terms of changes in function and structure in the biological organism.

The illness perspective deals instead with the perception the sufferer himself or herself has concerning the course of illness, on a past–present–future axis. Of central interest here are not biological changes, but the expected physical suffering and its implications in practical, psychological, and social respects. For purposes of distinction, the term "illness trajectory" is sometimes used as a designation for the illness perspective. Juliet Corbin and Anselm Strauss have defined the concept in the following way:

> "Course of illness" is both a commonsense and professional term. In contrast, "trajectory" refers not only to the physiological unfolding of a disease but to the total organization of work done over that course, plus the impact on those involved with that work and its organization. (And then the consequences of that impact for the work itself.) (Corbin and Strauss, 1988, p. 33)

The trajectory concept focuses on the active role people play in shaping the appearance and meaning of the course of illness. By considering "the total organization of work done over that course," Corbin and Strauss also want to stress that the care staff are not the only persons involved. The primary sufferers themselves, their family, and a large number of other persons and institutions are also comprised in the total work organization. For example, this work can involve efforts on the part of the social insurance office, neighbors, and friends, and efforts by patient associations and other voluntary organizations. Since this definition depicts illness as something managed by many actors in interaction with one another, we should be able to talk about a sociological or social-psychological definition of the course of illness.

4. Uncertainty and Its Consequences

In a critical review of the sociology of illness, Peter Conrad (1990) calls attention to a source of suffering in chronic illness not present in the conceptual frameworks of Charmaz and Bury: uncertainty. Uncertainty is since a long time a central problem in clinical work. In particular, people have emphasized the difficulty in making correct diagnoses and the consequences this can involve for health-care staff and patients. But Conrad also discerns other forms of uncertainty that have been demonstrated in empirical research:

(1) Frequently, uncertainty arises when the first physical discomfort begins to appear. Is this a sign of illness or something totally normal and temporary? Many research results indicate that a strong tendency exists not to seek care until the situation becomes so difficult that all everyday routines have almost collapsed and a large spectrum of everyday explanations have been tested. Another element of uncertainty present during this initial phase concerns when problems can best be communicated to others. Should I tell my family now and risk worrying them unnecessarily, or is it better to wait and see how the situation develops? Such questions often worry the person who is starting to feel unwell.

(2) Another form of uncertainty arises when the problem has been defined as "probably medical." The person himself or herself and the family are in a state of uncertainty concerning what kind of an illness it is. Before a diagnosis has been made, the ill person and family experience great worry about the final verdict. In the spectrum of illnesses that pop up in their imagination, cancer is the most frightening. In those cases in which a diagnosis cannot be made immediately, the initial worry frequently becomes further intensified. The possibility that information about a serious illness will be withheld crosses the minds of many.

(3) When the diagnosis has been made, a third form of uncertainty arises concerning what is involved in living with the illness in everyday life, how the illness is going to develop, how the treatment will work, and the side effects of the treatment. No easy answers to these questions exist. Therefore, they often develop into lasting, central elements of how the illness is experienced. Not of the least concern is uncertainty about how the illness will develop, which is hard to bear and traps ill persons and their families in a state somewhere in between hope and despair.

(4) The fourth and fifth forms of uncertainty concern short- and long-term changes in physical well-being. Illnesses like multiple sclerosis, leukemia, and diabetes are initially characterized by alternate, relatively drawn-out periods of improvement and deterioration that place obstacles in the way of long-term life planning. With some other illnesses, like Parkinson's disease and rheumatoid arthritis, the body fluctuates from day to day or from hour to hour between an almost total lack of symptoms and severe or troublesome symptoms. The difficulty in predicting when the symptoms will strike next makes it more-or-less impossible to plan and organize everyday life. Many persons with severe chronic illnesses therefore come to live from day to day.

Bury (1988) mentions an additional form of uncertainty: that the ill person cannot know beforehand how others will react if he or she chooses to disclose information about the illness. Disclosure can be functionally significant and can lead to greater understanding on the part of other people. But disclosure also risks causing people to withdraw from further contact with or to discredit the ill person. Since reactions to news about illness are seldom communicated directly, frequently, the ill person does not even know afterward where he or she stands or how he or she is perceived. Without a clear awareness of the situation, ill people find it hard to know how they should act. Others and their attitudes largely shape the ill person's life situation—and by extension, his or her judgment of it.

Many of the forms of uncertainty mentioned here are replaced in time by an increasing certainty, although never to the point of absolute certainty. The most extreme consequence of Conrad's observations is that all attempts to define the essence of the illness experience inevitably distort to some extent its basic meaning. From the perspective of the sufferer, the meaning of the illness is never final but emerges from his or her interactions with the physical and social world. The meaning also changes with changes in the underlying biological state. Seen realistically, a good deal of the interactions with others and even some of the biological changes can be assimilated within an existing form, but Charmaz's three modes of experience also indicate that the subjective meaning of the illness can change qualitatively from one time to the next.

5. Discussion

I have presented and discussed two influential medical-sociological conceptualizations of chronic illness, namely "illness as a biographical disruption" and "illness as a loss of self-esteem." Theorists sometimes direct the objection toward both conceptions that they are not able to sufficiently distinguish aging from illness. They do not give a misleading picture of what it means to be chronically ill (irrespective of age), but some of the processes that the ill person goes through are also a part of normal aging. Even the healthy elderly person often must change his or her habits, loses earlier social contacts, is subjected to discrediting actions, is considered less worthy, has a poorer economic situation, and loses his or her previous independence. But what can seem like a weakness from one perspective can be seen as a strength from another perspective. As tools for understanding the elderly in elderly care, these conceptions benefit from this duality. Ill, elderly, or ill and elderly? With some changes, these theories can provide a more in-depth understanding of these human conditions.

Both of the basic conceptualizations—and the conceptual amplifications and clarifications I have made—promote greater understanding of what it means to have a prolonged and serious illness. They also give us the beginnings of a language through which experiences of illness can be communicated and discussed. In this way they complement the medical understanding of illness. They make the ill individual, not the disease, visible. In-depth knowledge is a goal in itself. Tolerance and genuine communication grow out of understanding. I could end here, but will return to the themes in the article's introduction by pointing out some practical consequences for elderly care, some of which may be obvious and already considered, but others of which have received less attention.

(1) Contracting a serious long-term illness almost always means a loss of self-esteem, as Charmaz (1982) points out. Elderly-care staff will likely try to be supportive by talking about "the good old times." Used with care, such communication can be good. A risk exists, however, that it will lock the receiver of care into continual grief over what has been lost. Perhaps any ill person will miss what she or he has had previously, but feeling grief means finding yourself in a state where you cannot see anything to replace what was lost. In order to help the sufferer build up his or her self-esteem, that person must be activated in a way that suits him or her. To identify the "right" activities, the caregiver must learn about the sufferer's hierarchy of values. The most suitable activity is the one highest up in the system of preferences that is at the same time manageable despite physical limitations. Subjecting the elderly and ill individual to activities that he or she cannot manage further worsens self-esteem. Subjecting a person to activities that are not at all associated with his or her preferences can be offensive.

(2) Physical discomfort that is lasting but for which no medical explanation has been determined causes worry and sometimes anxiety (Bury, 1982; Conrad, 1990). Care staff must therefore be constantly observant of complaints from the elderly concerning changes in their physical condition. If the complaints concern discomfort associated with the disease, that fact should be made apparent to the sufferer. If they concern other types of discomfort, then naturally this should be followed up medically as soon as possible.

(3) Someone who has recently fallen ill suffers from physical discomfort and from uncertainty about how the illness will develop (Conrad, 1990). Perhaps information about the illness and its treatment should be given to the elderly person and his or her family. Unfortunately, making sure that the patient and family have understood such information is not always a matter of course. Nor does it go without saying that new information is supplied when necessary. The family, in particular, can suffer from a lack of information and insight concerning the patient's illness. We need to clarify the medical picture for the elderly person and his or her family. They may find it awkward to ask and difficult to know what to ask about. The personnel should therefore readily give adequate and

understandable information; the elderly individual and his or her family should not have to ask for it.

(4) Suffering at the prospect of the future is due only in part to inadequate medical knowledge. The primary sufferer and family also experience uncertainty about how daily life will be affected, both in the short term and from a slightly longer view (Conrad, 1990; Bury, 1988). Care staff should map out previous life circumstances with the elderly person and his or her family and the parties should discuss what adaptations and changes may be necessary, not only in the present but also later in the course of the illness. Such "trajectory prognoses" need to be looked over and revised at regular intervals.

(5) Illnesses that fluctuate back and forth generate uncertainty about the coming day (Conrad, 1990). Plans for the day can be difficult to carry out if the disease has entered an acute phase. If ambitions have been at a low level, the body may allow for more than the ill person had initially counted on. In such cases, flexible care routines should take fluctuations in the illness into account. One way for the staff to handle the short-term uncertainty is to consciously prepare alternative plans. If the elderly person is also involved in this, much disappointment and frustration can be avoided.

(6) A person who works professionally in helping others easily comes to consider himself and herself and colleagues as the only ones who are working. The trajectory concept (Corbin and Strauss, 1988) notes that care is a collective activity in which the elderly person and family are deeply and continually involved. During the course of a normal day, the elderly person and his or her family can carry out a good deal of care-related work that goes more-or-less unnoticed. Not even the apparently one-sided achievement of giving the elderly person his or her medication can be carried out without the active effort of the elderly person. Staff members in elderly care need to develop their ability to identify the forms of work that are carried out by the elderly individual and family and take pains to show appreciation for this work.

(7) The patient's needs vary with the type of illness and with where the sufferer is along the course of the illness. We can divide the course into different periods. Using Kathy Charmaz's (1992) three modes of experience as an intellectual tool, we can draw the following conclusion: For the person who still regards his/her illness as "an interruption in everyday life," the need for medical knowledge can be great, to prepare for what is to come or to attain insight about the condition's status as non-temporary. In both cases, the patient needs information about what the alternatives to his or her present daily life could be like. For the person who sees the illness as a permanent "intrusion," the need for help and support for reorientation is often dominant. For the person who is "immersed in illness," existential questions and needs can instead come

to the foreground. Sometimes the "immersion" can have a mental rather than a physical cause. Needs can then be of different kinds, all the way from the need for medication to the need for kinship or self-fulfillment. This way of looking at the variation in needs is fairly rough and incomplete, but it provides a hint as to where to begin the work of identifying the individual's needs.

(8) As human beings, we all maintain a story about ourselves. Some put this down on paper in the form of a written autobiography, but many bear it in silence or describe only fragments of it in interaction with others. The story contains relevant elements of our life and stretches from the past up until the present, and then onward into an imagined future. What characterizes our self-story is that a common thread runs throughout it: One thing leads to the next. The story provides a feeling of identity, context, and direction. When a person contracts a serious, long-term illness that common thread is lost. Becoming ill is a life event not present in the earlier story that must in some way be incorporated into it. To regain the feeling of context two questions in particular must be answered: "How did I end up here?" and "What is the future going to be like?" Care staff talking with the elderly person about his or her earlier life in general is an event that often happens naturally. The theories I have presented here point out that it may be constructive to give the elderly individual the opportunity to talk about his or her illness experiences. Only in this way can his or her life be patched up again and be understandable. Through the story, the sufferer's feeling of estrangement and isolation can be broken through.

(9) Realizing the vision for care described above requires a care organization in which different staff members have special responsibility for and develop close, continual contact with individual elderly persons and their families.

It may appear that the above points overstep the framework of needs I presented as a base for care in the introduction of this article. But they do lie close to it. Many points deal with measures for reducing the uncertainty that serious illness creates. Uncertainty is the direct opposite of the security that Maslow (1954) considers a basic human need. Even the second-last point—that the elderly person and family should have the opportunity to talk about their experiences with respect to the illness—deals in part with reducing uncertainty. We all have a need to know who we are, where we come from, and where we are going. By relating his or her case history, the elderly individual can obtain increased clarity concerning those questions, which are perhaps the most basic of all with respect to the possibility of feeling a sense of security.

As I have suggested, the telling also creates a feeling of kinship with others. Several of the points deal with ways of strengthening self-esteem.

Translated into Maslow's hierarchy of needs, the needs for appreciation and self-fulfillment are in focus in those cases, rather than the need for security. Read in this way, the theories touched upon here do not fall outside the framework of needs. They can be considered rather as a means of filling the otherwise empty and meaningless framework with something concrete.

A final note: Many of the visions for care in the above list are naturally valid for the care of all persons with serious long-term illnesses. I have focused particular attention on elderly care because of the extreme loss of previous everyday life and social relationships that people obliged to live their lives within an institution will come up against. Biographical disruption and loss of self, which for a person who is still living relatively independent of society's support structure can best be considered as tendencies in his or her existence, are a definitive and irrevocable reality for a large number of persons who are elderly and seriously ill.

Part Five

CONSTRAINTS ON INTERVENTIONS: SCIENTIFIC AND ETHICAL REQUIREMENTS

Eleven

ALLOCATION OF SCARCE RESOURCES IN HEALTH CARE: VALUES AND CONCEPTS

Per-Erik Liss

1. Introduction

In many countries, a gap exists between the population's need for health care and available resources. These nations have attempted to eliminate or reduce the gap through such activities as improving efficiency and narrowing responsibilities. Since these measures have proven insufficient, we must make decisions regarding how to best use the scarce resources. No disagreement exists on this point, and the situation is not new; health care resources have always been scarce. What is new is that the situation of scarcity has been publicly acknowledged and that decisions on how to use the scarce resources, and the grounds for those decisions, have been made public. The decision-making process should be open, and may include the participation of citizens. The idea of citizen involvement is sometimes presented as empowerment of the public (Saltman, 1994).

The priority-setting and rationing processes involve key decisions in the sense that they have consequences for people's health and quality of life. Because the decisions sometimes concern life and death, we want them to be rational and based on solid grounds. This means that we want decisions to be based on facts, concerning, for instance, health care needs or the relative effectiveness of different treatments. But not every fact will do. The choice of facts should be rational. If, for instance, we set priorities in accordance with health care needs, facts about these needs are essential—but before we have a clear understanding of the concept of health care need, we do not know exactly which facts. Whether we should focus on health care needs or on other things depends on the ethical principles that guide the priority-decision. Health care needs only become interesting if the principle of need guides the priority-decision.

In other words, three types of issues are involved here: issues about facts, concepts, and values. A hierarchy exists between these categories. We do not know what facts to look for until we have an answer to the conceptual issue, and we do not know what concepts to use before we have an answer to the value issue.

Different views exist on what should constitute a value base for decisions concerning the allocation of health care. This essay will not include any attempt to give an overview of these views. Only one example of a value base will be used for illustration. The essay will mainly concern conceptual issues. It begins with an analysis of the concepts of allocation, prioritization, and rationing, and proceeds with a presentation of the value base that was implemented through the Swedish Health Services Act. A central principle in this value base is need. To understand the need principle, we must analyze the concept of health care need. This analysis leads to the concept of a goal and to the question of the goal of health care. As the goal of health care basically is a value issue, a value base that includes a principle of need must also include a principle for choosing the goal of health care.

2. The Concepts of Allocation, Priority-Setting, and Rationing

We can distinguish between two situations of scarcity in health care. In the first situation, we can meet all needs for health care but not at the same time; the needs must be met in a particular order. This situation is common and unavoidable. All patients in the waiting room, for instance, cannot be treated at the same time. In the second situation, we cannot meet all needs within a set time span. In principle, we can in such a situation choose between the following alternatives:

(1) meet no needs at all;
(2) meet every need but only to some extent;
(3) meet some needs optimally, and others not at all;
(4) meet some needs optimally, and others only to some extent;
(5) meet some needs only to some extent, and others not at all;
(6) meet some needs optimally, some needs to some extent, and others not at all;
(7) meet every need only to a partial extent, but some needs to a larger extent than others;
(8) meet some needs optimally, some needs to a large extent, and others to a small extent.

These activities are sometimes called "rationing" and sometimes "setting of priorities." As they involve the distribution of resources, the term "allocation" is also involved. Different people use the three terms in different meanings, under different circumstances, to cover one or more of the items. A consistent conceptual structure will reduce the risk for confusion.

A. The Concept of Allocation

We normally use the term "allocation" to refer to a process or a state. When we say, for instance, "We did not follow the right guidelines when we allocated the scarce resources," we refer to allocation as a process. On the other hand, when we say, "The allocation is not fair, because some groups now have too little," we refer to allocation as a state.

The term "process," in turn, may have two different senses. In one sense, we use the term to refer to a sequence of occasions that involve conscious decisions with a specific result in mind. Those who are making the decisions intend a particular result or state. I suggest the word "result-oriented" for denotation of this sense. In the other sense, the term refers to a sequence of occasions where the end-state is a result of chance or of several individual acts without coordination. In such a process, no intended result exists. The term "state-oriented" will here convey this sense.

We should not mix up what we refer to by the term "allocation" with what we do when we assess a process or a state. We can in some circumstances assess a state by assessing the process. If the process is fair, the state is fair, irrespective of how things are distributed. Or we may assess a process by assessing the state. If the state is fair, the process is fair.

Allocation as a result-oriented process means a conscious decision to divide and distribute something (for instance, resources) between two or more beneficiaries. Allocation of resources can occur between districts, hospitals, clinics, primary care centers, categories of patients, etc. The standard form of allocation is characterized by the decision to divide something between two or more recipients, distributed to each respectively.

An allocation implies an allocation of something, or that something is assessed when a result is considered. We can speak about allocation of resources of different kinds, but we can also speak about allocation of, for instance, diseases, injuries, or ill health. What we divide and distribute or what we assess when judging a result will be denoted as the allocation-category. When, for instance, we allocate some material resources between different clinics, the material resources constitute the allocation-category.

We may use the same allocation-category when carrying out the allocation as when assessing the end-state. But we can use different allocation-categories at different stages. Consider the following statement: "To change the allocation of resources between in-patient care and out-patient care will not lead to a change in the allocation of ill health in the population." "Resources" is here the allocation-category, when allocation is considered as a (result-oriented) process, while ill health is the allocation-category when allocation is considered as a state.

The allocation-category is different from the criteria used for assessing the process or state. The allocation-category concerns what is distributed, while the assessment criteria concern how it is distributed. The assessment-criteria relate to values—for instance, justice—and the choice of criteria is an expression of values.

However, deviations can occur from the standard form of allocation mentioned above with respect to two points. First, the term "allocation" may sometimes have a restricted sense, denoting only the decision to divide what constitutes the allocation-category. In some cases, we may need to describe the division only. In these cases, "allocation" is used only to denote the decision of how to divide the resources.

Second, the deviation concerns a substantial aspect. The standard form of allocation implies that all of the requests receive at least something. The term "divide" may have a wide sense, denoting also a decision with the consequence that a petitioner receives nothing at all. We can divide a given resource X between A and B in such a way that A receives everything and B nothing. What justifies our denoting such a division as allocation is that A and B have been considered in the process. To divide something between two petitioners, for instance, implies that both of them are considered.

This is a significant criterion: whether a result-oriented process should be called an allocation. When we give a coin to the beggar in front of us but not to the beggar around the corner, whose existence we have not realized, we are not making an allocation between the beggars. On the other hand, if we have two beggars in front of us and we give two coins to the one but nothing to the other, then we have allocated resources between the two beggars. We are here speaking of allocation as a result-oriented process.

But when we consider an allocation as an end-oriented process, the situation is different. The criteria of consideration no longer apply. A process where request A gets everything and request B gets nothing because B has not been considered may be called an allocation.

Allocation as a state can be a consequence of a result-oriented or an end-state-oriented process. In the case of an end-state-oriented process, the allocation need not involve intentions concerning the result (Buchanan, 1989). The allocation may be a result of pure chance or of several individual acts without coordination.

An allocation as a result-oriented process need not imply rationing or setting of priorities if resources are not scarce. But if that is not the case, and it probably never will be, rationing or setting of priorities may be part of the allocation process. The decision to divide a scarce resource between some petitioners may, for instance, imply that some petitioners get nothing (setting of priorities) or that every petitioner gets something, but less than is needed (rationing).

B. The Concept of Prioritization

To prioritize always means prioritization in accordance with something. We may, for instance, set priorities in health care in accordance with health care needs. But we may also prioritize between something; for instance, between treatments, patients, clinics, or health care centers. What we set our prioritization in accordance with is in some cases identical with what we prioritize between. In many cases, when we prioritize between health care needs, we do so in accordance with health care need. In other circumstances, we prioritize in accordance with something different from what we prioritize between. We may, for instance, prioritize between different groups of patients in accordance with health care needs. What we set priorities in accordance with will be called the prioritizing category. In those cases where we set priorities in accordance with health care needs, health care need constitutes the prioritizing category.

To set priority implies making a conscious decision or choice. Not every choice in health care, however, should count as a choice of priority. The priority-decision concerns a choice between considered alternatives. We cannot say that we have chosen alternative A over alternative B if we have never considered B; the alternatives must belong to the same prioritizing category. A decision to allocate a health care resource to a person in need but not to a person not in need should not count as a priority-setting decision. Likewise, to choose an effective treatment instead of a non-effective treatment should not be considered as a priority-setting decision. Non-effective treatments do not belong to the prioritizing category.

The prioritizing category, determined by the priority-setting principle, is an essential element in the concept of prioritization. For example, if the principle of need is guiding the setting of priority in health care, we should set priority in accordance with health care need, and health care need constitutes the prioritizing category.

Setting of priorities normally involves a combination of choices or decisions. Setting of priorities in accordance with needs means that we consciously create an order between the needs. This is one essential component. We are normally not content with ordering the needs, however. We want to take further measures. In the first situation of scarcity mentioned above, we could meet all needs, but not at the same time. Setting of priorities in such a situation includes the decision to rank needs in a particular order—and to optimally meet all of them in the order of ranking. In other words, a priority-decision may involve more than one choice and, therefore, a set of sub-decisions.

In the second situation of scarcity, all needs cannot be optimally met within a set time span. I mentioned above eight different ways of handling such a situation. Only two of them do not involve prioritization: items one and two. The process in item three involves three sub-decisions: how to rank the needs; where to draw the line in the order of ranking; and how to treat the needs on either side of that line. The three sub-decisions together constitute a priority-decision. We may compare this process with the process in item four, which includes sub-decisions concerning ranking and where to draw the line, in addition to a sub-decision concerning the extent to which needs below the line should be met. The first two decisions are sub-decisions concerning priority-setting, while the third decision—to what extent the needs below the line should be met—is not a priority-setting decision. This sub-decision, I will argue later, should be considered as rationing.

The reasoning above illustrates that priority-setting decisions concern the following kinds of sub-decisions: how to rank the needs, where to draw the line in the order of ranking, and whether we should at all optimally meet the needs on their respective side of the line. The sub-decision of how to rank needs is an essential component in priority-setting. A choice that does not include such a sub-decision should not be considered as priority-setting. The other two sub-decisions have only a contingent status. They may or may not be combined with the first kind of sub-decision in a priority-decision, and they can only be considered as sub-decisions concerning priority-setting in combination with the first sub-decision. This implies that we can have several kinds of priority-decisions, depending on what kinds of sub-decisions are involved.

One priority-decision involves a sub-decision to rank the needs and a sub-decision to meet all of them optimally in the order of ranking. This process is used in the first situation of scarcity, where we can meet all needs but not at the same time. A second priority-decision involves a sub-decision to rank the needs and a sub-decision to draw a line in the order of ranking. A third priority-decision involves a sub-decision to rank needs, a sub-decision to draw a line in the order of ranking, and a sub-decision to meet the needs on one side of the line optimally and the needs on the other side not at all. The sub-decisions are integrated in practice, and the order between them may sometimes be different. The third priority-decision can, for instance, start with the sub-decision not to meet all needs.

That reasoning leads to the following conclusion. In order for a decision concerning a choice between X and Y to count as a priority-setting decision, three conditions must be fulfilled:

(1) we must consider X and Y;
(2) X and Y must belong to the same priority category; and
(3) the decision must imply the creation of an order between X and Y.

The three conditions together constitute a sufficient condition for a decision to count as priority-setting. Besides, the concept of prioritization characteristically includes a value component. What constitutes the prioritizing category is normally connected to something that is positively valued. To set priority is in general considered as something negative. The alternatives that we must choose between all represent, in some sense, something positive. We usually do not consider it as setting of priority when we choose between only negative alternatives.

We may also make a distinction between active and passive prioritization. We have so far been speaking only of active prioritization: the making of conscious decisions in advance of implementation. Passive prioritization, on the other hand, concerns a decision to accept the end-state of an allocation process, implying that one instance is favored against another. But the acceptance must be voluntary in the sense that the person who is accepting has the power to change the end-state. A standard case of passive prioritization would be when a person external to the decision-making process decides not to attempt to make a change in the end-state. The process that produced the end-state may be result-oriented or end-state-oriented; it may imply that some petitioners will receive nothing while others receive what they need.

C. An Open Priority-Setting Process

A priority-setting process is open insofar as it is made public. Being public here means being available for those who want to know. The openness of the process can vary along two parameters: those to whom the decisions are available, and what is available.

The openness may vary in accordance with the number of persons to whom the decisions are available. We have a minimal degree of openness when the decision is available to only a few persons and a high degree of openness when it is available to most people.

The openness of the process may also vary with what information is available. Four factors are central here: the content of the decisions, the grounds for the decisions, the reasoning or compromises behind the decisions, and the consequences of the decisions. The content of a decision concerns information that a decision has been made and what has been decided (for example, that maternity care has been given a higher priority than rehabilitation care within a hospital).

Giving information about the grounds for the decision available can also increase the degree of openness; for instance, that the decision has been based on an assessment of health care needs or on a cost-efficiency analysis. Making information about the reasoning behind the decision available may further increase the degree of openness. That information may concern, for instance, the ranking of needs or how the consideration of needs is balanced against the consideration of cost-effectiveness. Making knowledge of expected consequences available can increase the openness. The consequences may concern economic issues or people's health.

D. Explicit Versus Implicit Prioritization

The distinction between explicit and implicit prioritization occurs along two dimensions—the degree of consciousness and whether we use guidelines or not. According to the first dimension, explicit prioritization means a conscious decision. The explicitness here concerns the grounds for the decision or the consequences of the decision. Awareness of the values that have determined the priority-decision signifies explicit prioritization. Consequently, not being aware of the grounds for the decision or the consequences of it signifies implicit prioritization. But we cannot speak of an implicit prioritization if the agent is unaware that a choice has been made between alternatives in the same prioritizing category, because by definition, the choice must be a conscious choice in order to count as a priority-setting decision.

The second dimension concerns the use of guidelines. Not using guidelines when making the choice signifies, according to this interpretation, implicit prioritization while the use of guidelines signifies explicit prioritization.

A priority-decision may be explicit and open. We may base this decision on some guidelines or principles; the decision-maker may be aware of the grounds for the decision; and the decision and the grounds for the decision may be made public.

E. The Concept of Rationing

An alternative way of dealing with a situation of scarcity where everybody cannot get what they want or need is to give a small portion to all; we will not optimally meet their needs. Intentional restriction is the essence of rationing. In the second situation of scarcity, where we cannot optimally meet all needs within a designated time span, rationing would be a possibility. The term "rationing" will in this essay be used to refer to a decision to not meet the need for health care optimally. Such decisions are included in the processes illustrated by items two, four, five, six, seven, and eight.

The meaning of "optimally" is crucial in the stipulated sense of "rationing" presented here. That we meet a need optimally can mean several things. It can mean that:

(1) the goal of need is achieved to an optimal level;
(2) the goal is achieved within a set time span;
(3) the goal is achieved with some degree of security;
(4) the goal is achieved with few secondary effects; or
(5) the goal is achieved at a reasonable price.

Rationing of health care in accordance with need can thus meet one or more of these five conditions. We can, for instance, ration health care by extending the waiting-time for treatment or by using cheaper drugs with more secondary effects. We can also ration health care by setting a high price on some treatments, with the consequence that not all people can afford those treatments (Beauchamp and Childress, 2001). What we primarily think of when we speak about rationing is, perhaps, restriction along the first dimension. In such cases, we will not optimally achieve the goal of need (the phrases "the goal of need" and "optimally achieve" will be explained when we deal with the concept of health care need). To restrict rehabilitation activities would be one example of not meeting a need for health care optimally.

We can combine priority-setting and rationing in the sense that the process of rationing involves priority-decisions, a point that items four through eight illustrate. A rationing process involving the decision to meet some needs optimally and others only to a partial extent (item four) may be preceded by the priority-decisions of how to rank needs and where to draw the line.

We can distinguish between two main kinds of rationing processes. The first deals with pure rationing, where no priority-decisions are involved. This process includes the decision not to meet all of the needs optimally and the decision concerning the extent to which we should meet the needs (item two). The second kind of rationing process involves priority-decisions. These decisions concern how to rank needs and where to draw the line (items four through eight).

We can consider priority-setting and rationing on more than one level. So far, we have been concerned with the individual level. We decide on and rank the individuals' needs for health care. But we can also speak of a population's need for health care, which then means the sum of the individuals' needs. A process of setting priorities on this level may mean that we do not at all meet some needs in the population. Another way to describe this process is to say that the population's needs have not been optimally met, and not meeting needs optimally is a kind of rationing. In other words, a process that is considered as priority-setting on one level may be considered as rationing on another level.

3. The Value Issue: An Example of a Value Base

The choice of a value base for decisions of allocation is crucial. The value base determines the answers to subsequent issues concerning concepts and facts. In addition, it determines what other issues will become interesting. To illustrate the significance of the choice of value base, I will in this essay use an example from Sweden.

In 1992, the Swedish government decided to appoint a parliamentary commission with the task of discussing the role of health services in the welfare society and to highlight the basic ethical principles by which prioritization ought to be guided. The findings of the commission on the setting of priorities are now incorporated in the Swedish Health Services Act (*Priorities in Health Care*, 1995).

The value platform for setting of priorities in health care in this act contains three ethical principles: the principle of human dignity, the principle of need and solidarity, and the principle of cost-efficiency.

The principle of human dignity is based on an idea of universal human equality. According to this principle, human beings have equal worth and an equal right to respect and dignity. Human dignity does not depend on people's personal qualities or functions in the community, such as age, ability, social status, income, health status, etc.

The principle of need and solidarity asserts that resources should first be committed to those fields, activities, or individuals where needs are greatest. This principle implies that we should distribute most of the health care resources to those who are most in need. The greatness of need should then be assessed in terms of the severity of the injury or illness.

Solidarity means, in part, an effort to equalize the outcome of care as far as possible, at the same time devoting special consideration to the needs of the weakest. Some people are not aware of their human dignity or have less chance than others to make their voices heard; children, senile patients, the unconscious, and the confused, for example, or persons with serious mental disorders. As a risk exists that people will not communicate these needs, the idea of solidarity demands particular attention.

The principle of cost-efficiency means that we should try to get as much good as possible out of the scarce resources in an individual case. This may imply, for instance, that we should choose the most cost-efficient treatment when we have a choice. According to the commission, however, we should not be dogmatic here—we should "aim for a reasonable relation between cost and effect."

The report delineates which other principles we are not allowed to consider in this context. The commission rejects the use of the utility-principle—the idea that scarce resources should be used in a way that maximizes the utility of society. It also rejects the principles of lottery, demand, and autonomy. Nor are we allowed to consider things like age, self-inflicted injuries, economic ability, or social functioning. The commission emphasizes the importance of paying particular attention to the weakest people, those who cannot themselves demand health care.

Critics have turned against the ranking-order of the principles in that platform (Williams, 1996). The commission ranked the principle of need before the principle of cost-efficiency in a lexical order. This means that the size of the need should determine who we should treat when we cannot treat all of them. The size of the need should be determined by the severity of the illness or injury.

This ranking-order of the principles appears quite reasonable in most circumstances. Normally, we prefer giving the scarce resources to those who are most ill than to give them to those who can benefit most from the resources, when these two considerations do not coincide. The ranking-order can also have the effect that a seriously ill person, for whom we can do little, is ranked before a slightly less ill person whom we can restore to a normal life.

We normally consider this outcome unacceptable. The absolute character of the ranking-order, which does not allow exceptions, is unfortunate.

The three principles have different guiding functions in the value base. The principle of human dignity is a restrictive principle that constitutes a framework beyond which we are not allowed to pass, and tells us what should not guide our decisions. The other two principles are active ones: They tell us what should guide our decisions. But to give guidance, the principles must be interpreted. We will in this essay concentrate on the principle of need. The crucial thing for a reasonable interpretation is the analysis of the concept of health care need.

4. A Conceptual Issue: The Concept of Health Care Need

A. A Teleological Need

According to one academic tradition a need is something instrumental or teleological. We are then speaking of a teleological need (Liss, 1993). The existence of a need, and what is needed, is here related to a particular goal. We need food, for instance, in order to survive, and we need love in order to live well. The statement "P needs X if X is necessary for realizing a particular goal" illustrates this view. That something is necessary for realizing a goal implies that something is lacking; need is a deficiency of some kind. To put it otherwise, when a need exists, a gap also exists between what is and what should be. Assume a situation where Mr. Smith is said to have a need for a wheelchair. We typically use this locution when Mr. Smith lacks the wheelchair (what is) and the wheelchair is necessary in order for Mr. Smith to live a decent life (what should be).

B. Need as an Object

We can also find another way of using the term "need." We can say that a wheelchair is a need of Mr. Smith's, or that food or love is a human need. Need here refers to the object needed—normally, to some kind of treatment or care. For example, in a situation where the patient feels great discomfort, a gap exists between what is—the patient feeling discomfort—and what should be—the patient feeling comfort. This gap constitutes a need. Insofar as health care is necessary in order to reduce or eliminate the gap, health care is the object of need. We can say that two needs are involved here—need as a gap and need as necessary treatment for eliminating the gap.

C. The Teleological Need: Need as a Gap

The adherents of the teleological tradition, who consider need as a gap, specify the concept differently. They all take the view that need is something goal-related: Care is needed insofar as the care is necessary in order to realize a goal (Wiggins, 1987). Two criteria must be fulfilled before we can say that a person has a need for a particular treatment:

(1) A gap exists between the actual state of the patient and a goal; and
(2) The treatment is necessary for realizing the goal.

According to this definition, a need occurs when condition one is fulfilled, and a need for a particular treatment occurs when conditions one and two are fulfilled. The term "need" here refers to a gap or state of deficiency. An example: After going through a treatment, the patient feels fairly anxious. This situation is not good; patients should not be anxious. In other words, a gap exists in this situation between the actual state of the person—anxiety—and the goal: freedom from anxiety. Criterion one is fulfilled, which means that a need exists. We can call this need a health-related need. If some kind of treatment is necessary in order to fill the gap—to eliminate the anxiety—then criterion two is fulfilled and a need for health care exists.

The goal component plays a central role in the concept (Barry, 1965). First, the goal determines the object of need (the thing needed). The goal "to live as a farmer" generates a set of general life needs that partly differs from the set generated by the goal "to live as a nurse in the city." Different goals generate different sets of needs. A clearly defined goal is therefore a prerequisite for a reasonable assessment of needs.

Second, the goal is the justifying component in the need concept. Claims of need have a special force in most cultures. To say that a person needs a treatment exercises a greater force than to say that the person wants the treatment—not because the treatment in itself is considered valuable, but because the treatment is necessary in order to reach a goal that is considered valuable. Claims of need get their force from the desirability or value of the goal. This explains why we normally consider it more essential to meet the need for a meal (in order to survive) than the need for ice cream (in order to get refreshment).

When the goal is a moral value, the goal of need becomes a source of moral obligation. We can then argue that a *prima facie* obligation exists to satisfy the basic needs of people, or that people have a *prima facie* right to satisfy their basic needs. The basic assumption behind these arguments is that the goal of need, not the object of need (the thing needed), generates the moral obligation.

The teleological view of the general concept of need may be summarized in the following formula: A need exists when the goal is not realized, and a need of a particular thing exists when that thing is necessary for realizing the goal. We have here a general formula that we may use in characterizing not only a need for health care, but also every other kind of need: We need a particular thing or activity insofar as it is necessary for realizing a goal.

I have so far discussed only the general concept of need and thereby pointed out the importance of the goal. The goal we are concerned about here is the state we want to achieve by meeting the need. When we speak about a health care need, we presuppose a goal to be achieved—a goal of health care. What is then the goal of health care? This is a fundamental issue when we try to assess the need for health care. Health is an often-suggested goal. We will not, however, go into this particular issue; the emphasis will be on the concept of a goal. Other articles in this volume deal with the concept of health (see also Nordenfelt, 2001).

5. The Concept of a Goal

We have to address a few fundamental issues concerning the conditions for the setting of goals. One issue concerns the concept of a goal: What do we refer to when we are using the term "goal?" Do we, for instance, refer to something that is wanted, or do we refer to something that will be achieved? When we speak about the goal of health care, do we then mean a goal that health professionals choose or should choose?

A second issue concerns the different components of a goal-structure. A tendency exists, when discussing goals in general, to consider a goal as an altogether single entity. This may be acceptable in some circumstances, but it is quite unsatisfactory when discussing the goal(s) of a complex enterprise like health care. In such a case not just one single goal exists, but a structure of goals normally organized in a hierarchy.

A third issue concerns the most important dimensions of a goal. The setting of a goal implies the making of decisions based on three dimensions—quality, quantity, and temporal considerations. The quality dimension concerns what is to be attained; the quantity dimension concerns how much of the specified quality is to be attained; and the temporal dimension concerns when the goal is to be attained.

All these aspects and dimensions of a goal can be combined in different ways. A decision concerning which combinations are of interest is vital to the discussion of health care or its goals. Exploring these issues can help us create a conceptual platform for a discussion of the goals of health care.

The term "goal" is often used to designate a state considered to be an end of a sequence of events. The events may have actions or natural processes as their causes. For a state to count as a goal, it must be related in some way to an attainment. It must be conceived before its attainment: We must have anticipated or determined the goal in advance. This is possible in different ways, and it can be helpful to distinguish between different types of goals (Liss, 1996).

A goal can be a state that someone wants or wishes to attain. Most people set some goals in life concerning things like family, education, and occupation. But goals are also set by groups of people, for instance different professionals. We can also find a desired goal in health policy documents of different kinds. "[P]ublic participation in health work," a regional policy document in Sweden suggests, should constitute the goal for health-promotion in the region. According to the Swedish Health Services Act, "good health for the whole population" should be the goal of the health care system. Both these goals are politically chosen; they may be called desired goals.

But a goal may also be something that ought to be attained. "Ought" can be used in different senses: ethical, legal, rational, etc. This type of goal may be called a normative goal.

A. Goal Versus Ideal

We cannot attain everything that we want or wish to attain. To call something that is wanted but cannot plausibly be attained an ideal is a reasonable idea. Plausibility then has a key function within this idea, serving as a sort of line of demarcation between what end-state can plausibly be considered as a goal and what end-state cannot. We have a goal when we can attain the end-state with some degree of plausibility, and we have an ideal when we can attain the end-state with a low degree of plausibility. When the plausibility of attaining the end-state is extremely low or nonexistent, we have a utopian ideal or a vision. The World Health Organization's often-cited definition of the concept of health is a good example of an ideal: "Health is a state of complete physical, mental and social well-being and not merely the absence of disease or infirmity."

B. A Goal-Structure as a Hierarchy

We can try to attain a goal for different reasons. We may try to attain a goal for its sake, or for the sake of another goal. Sometimes a longer sequence of goals exists. We can organize this sequence as a hierarchy. A goal we try to attain for its sake will be called a basic goal; a goal we try to attain for the sake of another goal will be called an instrumental goal.

We can illustrate the relation between instrumental and basic goals by saying that we achieve a basic goal by achieving the instrumental goal. For example, we can improve a patient's quality of life by comforting him or her, and we can comfort the patient by sitting by the bed and talking to him or her. Talking to the person and giving comfort are here instrumental goals for the basic goal of quality of life.

Different kinds of instrumental goals exist. Two kinds are of particular interest here: operational goals and sub-goals. The difference between these two instrumental goals is that when the agent has attained an operational goal, he or she need not perform any further action in order to attain the basic goal. In the example just mentioned, sitting by the bed and talking to the person gives comfort that in turn improves the quality of life. Comforting the person is here an operational goal for the basic goal of improvement in quality of life.

The other type of instrumental goal is a sub-goal. Such a goal is only a step you take toward the basic goal; further steps are needed from you in order to reach the basic goal. For example, to be able to talk to the person, you must get in contact with him or her. You may have to sit by the bed. Sitting by the bed is then a sub-goal for the basic goal of improvement in quality of life, because it is insufficient in this case for improving quality of life. You have to perform further actions—you must also talk to the patient.

C. The Operational Goal

When the agent has attained an operational goal, he or she need not carry out any further action to attain the basic goal. But other people may have to carry our further actions for this result to obtain. We therefore have two kinds of operational goals. Attaining the first kind of operational goal means that no further action is needed on the part of anyone to attain the basic goal. The intake of a painkiller is an example of this kind of operational goal: No further action is needed for attaining alleviation when this operational goal is attained.

Attaining the second kind of operational goal means that further action is needed on the part of some other agent than the original agent. Perhaps most operational goals in health care are of the second kind. Restoring health, for instance, normally involves several professions, each of which has its operational goal. In such cases, the attainment of only one operational goal is not sufficient for attaining the basic goal; the operational goal of every profession involved must be attained. A particular profession may in some circumstances have more than one operational goal. These goals are in such circumstances organized in a sequence. Suppose that a sequence consists of a basic goal and two operational goals. The agent then attains the basic goal by attaining the operational goal closest to the basic goal. He or she realizes this by attaining the first operational goal in the sequence. It is sufficient for the agent to attain the first goal in the sequence in order to attain its part of the basic goal.

D. The Sub-Goal

The characteristic of an operational goal, then, is that when the agent attains this goal, no further action is needed on his or her part in order to attain the basic goal. Sometimes, however, it is insufficient for an agent to attain a state in order to attain the basic goal: Further action is needed on the part of the agent. Such a state will be called a sub-goal. Buying a painkiller at the chemist's is an example of a sub-goal, as further action is required of the agent in order to attain alleviation.

E. The Domain of the Enterprise

An action-plan may direct the process of curing and caring. Such a plan may involve sub-goals and operational goals in addition to the basic goal. We can describe the processes in terms of the goal-structure. To provide a complete description, a further item in the goal-structure must be considered—the domain of the enterprise.

A domain is related in a way to a goal-structure. It concerns the type of action to be employed in achieving the goal. A simple way of describing the relation between domain and goal-structure is to take an example from a country's defense system. The air force, the army and the navy all have the same basic goal—a free country. But they have different domains and different operational goals. The air force will defend the country by actions performed in the air, the army by actions performed on the ground and the navy by actions performed at sea. Their special competence determines which area it is rational to operate in.

The domains of different professionals involved in health work, broadly seen, may be determined, in the same way as in the example above, by a rational application of their competence. The politician, the health-promotion officer, and professionals at the hospital, to name three, may all have the quality of life of the individual as the basic goal of their actions. But as their competences differ, it is rational for them to focus on different aspects of this goal. For example, the politician may reduce (or create conditions for a reduction of) unemployment; the health-promotion officer may educate people about health risks; and the health care professionals may cure diseases. Although the different professions have the same basic goal, they work in different domains and therefore have different operational goals.

F. The Goal-Structure Exemplified

Consider the following items as possible goals of health care: the controlling of disease, absence of disease, health, and quality of life. Insofar as the concepts of disease, health, and quality of life are defined in a reasonable way, they denote different states. When health, for instance, is defined in terms of ability to realize vital goals, the state denoted by "health" is different from the state denoted by "absence of disease." And if the phrase "quality of life" denotes a state that constitutes or is part of the good life, health is not identical to quality of life, but a condition for it.

Three of the items, although they involve different states, can simultaneously be considered as goals of health care, provided that they are parts of a goal-structure. Quality of life is then the basic goal of health care, while health is an operational goal. By attaining the health of the individual, health care has done its part for the individual to attain quality of life. If the individual cannot attain a reasonable degree of quality of life on his or her own, other professions or institutions may provide help. Health care attains the operational goal of health, in turn, by attaining the other operational goal—absence of disease. In those circumstances where a contribution from another enterprise is needed to attain health, medicine has done its part by attaining the operational goal of absence of disease.

What is the role of "the controlling of disease" mentioned above? The role is not the one of a goal of health care, but falls within the domain of health care because it concerns how the operational goals are attained. In the simplified sketch above, health care attains the operational goals by fighting diseases. In other words, the domain of health care, as exemplified here, is disease control.

G. What, How Much, and When?

Irrespective of which place a goal has in a goal-structure, it can be considered in three dimensions—the quality, quantity, and temporal dimensions. The quality dimension concerns what is to be attained. This dimension has dominated in the descriptions of the goals in the examples above. When we have placed absence of disease, health, and quality of life in a goal-structure, we have only presented the quality dimension of the goal. The quality dimension is the primary one because the other two dimensions cannot be determined before we settle the question of what is to be attained.

We can combine the three dimensions in different ways. The quality dimension is the primary one, and we may keep an entity within this dimension constant in relation to the other dimensions. But a specification on the quantity dimension may influence the specification on the temporal dimension, and vice versa. For example, if we want to attain a goal and if we specify the temporal dimension, we must then specify the quantity dimension in such a way that the goal can plausibly be attained. If a specified amount of a state cannot plausibly be attained within a specified time, this specified state is not a goal; it is an ideal.

H. The Goal of Health Care

Let us return to the issue of health carc need. A health care need is a particular need. According to the basic formula of need, health care is needed insofar as it is necessary for realizing a goal. The crucial question is then what goal would be realized by satisfying a health care need.

This goal cannot be discovered. The goal has to be chosen; a decision has to be made. But the decision need not be arbitrary. It may be based on a rational approach related to the purpose of the whole health care system. For instance, if good health is the goal of the health care system and if we should determine the system only by the needs of the population, it would be quite irrational not to choose health as the goal.

But health is not significant in itself. Health is significant because it contributes to people's quality of life. I therefore suggest that quality of life is the basic goal of health care, and health is then an operational goal of health care. But to achieve this goal, it may be necessary to achieve other goals. These other goals would then, in turn, be operational goals for health.

6. The Concept of Health Care Need

We can now approach the concept of need for health care. According to the general formula, a need for health care exists when:

(1) a gap exists between a person's actual state and a goal of health care; and
(2) health care is necessary in order to eliminate the gap.

This leads to a model for assessment of need for health care (Liss, 1993). The concept of need has three fundamental components: the actual state, the goal of need, and the object of need. Information about these components is necessary and sufficient for an assessment of a need. The primary element of this model, then, is obtaining this information. We may perform this task in three stages:

(1) Determining the actual state;
(2) Determining the goal of need;
(3) Determining the object of need.

While we might reverse the order of the first two steps, we cannot carry out the third step before the first and the second are finished.

A. The Goal of Health Care Need

We cannot always reach full achievement of a goal. A decision is therefore necessary concerning what degree of achievement should constitute the goal of health care need. That some state of achievement constitutes the goal of need means that when this state is realized, the need is satisfied, and no further action is required. The state of achievement that constitutes the goal of need will be called an optimal state. To meet a need for health care optimally means achievement of this state.

A need is basically a gap between the actual state and a goal. Satisfying a need therefore means the elimination of this gap. We have eliminated the need when we have eliminated the gap.

We can eliminate a need in more than one way. We can eliminate the gap by manipulating the actual state, changing the goal, or both. Our first thought is to manipulate the actual state—to give nursing care. In principle, we can eliminate a need by choosing a low state of satisfaction of needs. We can then, but this is hardly acceptable, dramatically cut down the individual's need for health care by using a concept of need that mirrors low ambitions with respect to the goal. But we can also do just the opposite. High ambitions could be established with respect to the goal, which would mean an increased amount of need for health care.

How should we determine the appropriate degree of achievement? No algorithm exists for this. The goal has to be chosen, and the choice is not only a matter of empirical investigation but also—and mainly—a matter of evaluation. The choice may be dependent upon some basic evaluation on the part of the entity that does the choosing. This implies that an issue at the second stage may be considered key for the whole assessment procedure. Who will have a dominating influence on the decisions? Whose values should determine the choice of the goal for health care need?

B. The Object of Health Care Need

Determining the object of health care need at the third stage concerns the services needed—the level and type of health care necessary for realizing the goal. The central question at this stage is whether health care is necessary to eliminate the gap between the actual state and the goal. Answering this question requires the competence of professionals. The determination at this stage is to a great extent an issue for health professionals, but not exclusively.

The situation varies with the number of treatments available for a specific purpose. Only one treatment may be available, or none at all. In many cases, however, a choice exists, including the option of renouncing treatment altogether. In making such a choice, professional considerations are not the only valid ones. Other values, in particular ethical ones, must play a central part.

C. Need Assessment and Economic Considerations

Another interesting, and more general, issue concerns what factors we should allow to influence the decision concerning the goal. Economic considerations constitute one of the most interesting factors here; the level of knowledge in health care is another one. We are tempted to say that economic factors should influence the decision. For instance, if an individual or entity can afford only a partial degree of achievement of the goal, it may appear pointless to choose a higher degree as the goal of health care need.

But to adopt this attitude would be to make a mistake. We must here distinguish between (1) the circumstance that a need exists, and (2) the choice of whether this need can or should be met. We err when we argue that a need exists only if it can or should be satisfied; the second condition should not be put as a necessary condition for the first. We make no contradiction when we say, "A need (for health care) exists, but this need cannot be satisfied." A thirsty person in the desert needs water even if there is no water within reach, and a starving person needs food even if he or she cannot afford it.

The existence of a need is altogether independent of whether we can afford to meet the need. Economic considerations should not therefore influence the identification of needs. They may only influence the decision as to whether we should satisfy the need, or with what treatment we should satisfy it.

7. Conclusion

If we adopt an ethical platform as a guideline for setting of priorities in health care, some concepts come into focus. In our example, the concept of health care need became more interesting in this light. The analysis of this concept led to the concept of a goal and to the argument that an understanding of the goal of health care need is necessary for an understanding of the concept of health care need. The goal of health care need, in turn, must be related to the goal of health care if the principle of need is to function as a guideline for setting of priorities in health care. This means that we cannot adequately analyze the concept of health care need until we have a clear idea of the goal(s) of health care.

Insofar as we want ethics to guide our decisions concerning setting of priorities in health care, ethics should guide our choice of the goal of health care. The approach I suggest here is built on the following idea: Human beings have some duties toward each other. A variety of actions may fulfill these duties. Some are quite simple and require only the effort of one person, while others are more complex and require the cooperation of several persons. Such cooperation is sometimes manifested in enterprises. The purpose of creating an enterprise may be based, then, on some duties toward human beings. The goal of the enterprise is to contribute to the fulfillment of these duties. The purpose of creating the enterprise determines its goal.

We must complement an ethical platform like the example we have been discussing with an ethical principle for the choice of the goal of health care. None of the three principles considered here can fulfill this task. This analysis also illustrates the necessity of interaction between an ethical analysis and a conceptual analysis. In this case, the suggested ethical platform leads to concepts that, upon analysis, lead us back to a revision of the ethical platform.

Twelve

QUACKERY

Ingemar Nordin

quack – *n.* An ignorant pretender to medical or surgical skill; one who boasts to have a knowledge of wonderful remedies; an empiric or impostor in medicine.—*The Oxford English Dictionary*
v. To treat after the fashion of a quack; to administer quack medicine; to seek to remedy or put right by empirical or ignorant treatment.—*The Oxford English Dictionary*
adj. 1. Characterized by pretentious claims with little or no foundation. 2. Dishonestly claiming to effect a cure.—*Webster's New World Dictionary*.

The term "quackery" is a designation meant to degrade therapies that we consider unscientific, non-medical, unprofessional, harmful, or just bad. The problem is that no clear distinction exists between quackery and true medicine when considering many borderline cases. For example, what is the difference between medical maltreatment and quackery or between self-medication and quackery? If we use the term only to depreciate any therapy that we dislike, we do not have much to discuss. If, however, the term stands for something that could help us to get a clearer view of medicine proper, or of alternative medicine, then it will be worth the trouble to analyze it further. I will not enter into a discussion about the ethics of quackery, or about how best to handle quackery in reference to law. The only ethical point of departure is that quackery is something bad, a kind of fraud.

1. Functionality

One possibility is to define quackery—as distinct from true medicine—as an enterprise not based on science. However, this definition is problematic. In medicine and in technology in general, we can find many cases where we have no detailed scientific explanation as to why the technique is working. For example, many drugs exist that have been used for decades to good effect, but where we can say neither that they are based on science nor that we have any good explanation as to why they work. And "new" therapies exist—like acupuncture, which has a documented effect against pain—that are not scientific in a strict sense. It would be strange to condemn such working therapies as quackery just because they do not have a scientific basis.

What if a doctor falsely claims that such a therapy has a scientific basis? If Dr. X claims that the medical treatment s/he recommends is based on the best available scientific theories of the day, but where this is not the case, would not Dr. X be a fraudulent quack? However, just being wrong about something may not be too bad. Dr. X may be mistaken about the existence of a scientific explanation of the functionality of the drug. Yet the drug may work perfectly and help the patient as promised. The doctor may have other good reasons to recommend the drug than a full-fledged scientific theory. For example, Dr. X could successfully have used it on other patients many times before. So not all incorrect claims about a therapy make it quackery.

Another option is to define quackery as non-professional medical activity. If people who do not have a formal education in medicine, or who do not hold medical credentials recognized and legitimized by the authorities, practice healing, then it would be quackery. The problem here is that the criterion puts most self-medication, treatment within a family and all so-called alternative medicine in the quackery box. All those things may well have good effects on the health of the patients, and may be both cheaper and more convenient than any professional help. Not all non-professional healing has to be inefficient and unworthy. Also, this criterion does not qualify ignorant, inefficient, and harmful medicine as quackery as long as people who happen to be professionals practice it.

A criterion often used for defining quackery in legal contexts is whether the therapy causes harm or not. However, since it happens that genuine medicine also may cause more harm than good, we have to be careful not to equate all medical malpractice or failure with quackery. The difference appears to lie in the justification for the therapy in question. If Dr. X correctly claims that good reason exists to recommend a therapy, then the doctor's use of that therapy is not quackery even if the treatment happens to go wrong and cause harm; the incident could be a single case of malpractice or bad luck. (If Dr. X has no rational grounds to recommend the treatment, however, then it may well be a case of quackery.) While the harm-criterion is too strong and broad in that example, it can also be too weak. Quack therapies such as homeopathy and the use of magnetic belts do not, by themselves, cause any harm; but they are pseudo-therapies if Dr. X falsely claims that they have been proven to work.

A better alternative, I think, is to base a definition of quackery not on harm, but on functionality. If proponents claim that a therapy works and that it has worked in tests, but this is not so, then the enterprise promoting and using the technique in question is quackery. With a definition along these lines we could, for example, include some uses of acupuncture among the standard medical therapies while dismissing others like homeopathy or magnetic belts as quackery.

A typical trait of quackery, and of pseudo-technology in general, is that the person practicing it makes untrue claims. It does not have to be a conscious lie; many quacks sincerely believe in their methods. But the person has to make a claim that is false, and the central falsity that would render a therapy quackery is to claim that it is genuine and working when it is not. This is how functionality comes into focus.

To be functional, a therapy, T, has to produce a wanted result, E. No drug or treatment could be expected to cure every occurrence or for every patient. But it should have an efficiency that is greater than zero or than what nature or chance is doing by itself. When testing a new drug, for example, you conduct a double-blind test. One group of patients is given the drug and another group—the control group—is unknowingly given a placebo. You also formulate two hypotheses, h_0 and h_1. Hypothesis h_1 claims that the drug will work to some degree; that is, that the patients in the test group will be cured to a higher degree than the patients in the control group. The h_0 is the so-called zero hypothesis which asserts that the drug will give rise to no result, that is, that there will be no difference between the test group and the control group. In this simple case, if h_0 is proven false, then h_1 is corroborated.

This testing process is not, however, sufficient for proving that T(E) works, or that it will work to a sufficient degree on a given patient in the future. The problem of induction here is unavoidable. We may have executed the tests in a situation, S, that may not be exactly the same when the patient undergoes the therapy. Also, because subjects are tested in groups, no guarantees exist that the patient who meets the practicing doctor belongs to the category of people who were cured in the test. Finally, even if we knew that the situation is the same and that the patient is similar to the people who were cured, we have no guarantee that the world is deterministic. The best we can say is that the tests show that T(E) will possibly cause E.

I should be more specific about the way in which T(E) has shown its "possibility." Given that we know nothing at all about T(E), anything is possible from a logical point of view. A tested technique, however, proves more than just a logical possibility; it also shows that in the test-situation, S, E was obtained when T(E) was applied, while E was not obtained in S when T(E) was not applied. If T(E) works, then situations exist in which E is obtained if, and only if, T(E) is applied. If the result of the test is positive—that is, if E is obtained precisely when T(E) is applied—then we cannot draw the conclusion that it will work every time, or to a sufficient degree, but we have corroborated the assumption that T(E) works. In spite of a serious test, we have failed to show that T(E) does not work. It appears, then, that T(E) in that sense has proved a worthy candidate for further trials and, possibly, for use. We have good reason to prefer a therapy that has a corroborated possibility of working, compared to others that have failed or that we have not tested.

I will take a "genuine technique" to be a treatment that has a corroborated possibility of working. Such a technique does not have to be based on a scientific theory, only to have been tested and found sufficient with respect to its expected result. Furthermore, a genuine technique does not have to work all the time; it could be a technique that has stopped functioning or that malfunctions. For example, even if my car does not start one cold morning, the technique is still genuine; it just does not work for the moment.

To demand that we have proven a technique to be working to be genuine is too stringent. The problem of induction makes that impossible, but we can demand that we have corroborated a genuine technique at some time and that it is possible that it works. If we accept these criteria, then the role of positive evidence in medical technology will not only be epistemological. The testing, and the positive outcome of this testing, proves that the therapy is genuine. In cases where we have tested new methods and drugs, the tests and the positive outcomes of those tests also make them genuine medical techniques. In medical technology, where the aim is to develop new techniques that can be used for solving health problems, the role of positive evidence is not only to clarify the epistemological question of whether therapies work, but also to contribute to the creation of such therapies.

Is the lack of a corroborated possibility—together with a false claim that such evidence exists—a sufficient condition to label a therapy as pseudo-therapy? What if T(E) has not been tested, but happens to work? Suppose that Dr. X tells a patient that the mixture T(E) will cure her illness, and that the doctor's claims concerning tests and corroborations of it are all wrong, yet T(E) does work and the patient gets cured. This is a borderline case. T(E) is not a genuine technique, but I would suggest that Dr. X is not performing quackery either, since I find it counter-intuitive to talk about pseudo-therapies that work. No pseudo-therapy ever works. Dr. X has not told the truth about some matters. Still, the doctor's claim about the functionality of T(E) is true, and the patient is cured by it as promised. In this case, Dr. X's therapy is neither pseudo nor genuine.

To summarize: Quackery consists in recommending and delivering therapies that neither work nor have a corroborated possibility to work, as if they were genuine and working.

The fraudulent act must include an implicit or explicit claim that the therapy is genuine and works. Without that, the practitioners have made no untrue claims. For example, suppose that Dr. X offers the patient a drug and says: "Here is an untried drug. As far as I know, it is completely harmless. However, I have no idea whether it will help you with your problems. But since I have nothing else to offer, you might just as well try it." Everything Dr. X says is true, and the doctor has promised nothing about the healing powers of the drug. The only questionable aspect of the act is that the doctor has no tested cure to offer, but that is not, in itself, sufficient for being quackery.

2. Usefulness

Corroborated functionality is not the only component of a pragmatic choice situation. The ultimate choice has to find a balance between functionality and usefulness. Therefore, not all the rational choices will be in accord with what we have corroborated the most. If, for example, the drug D_1 is well corroborated but has some nasty side effects, while the drug D_2 is corroborated, but less so and with small side effects, and if the illness is not acute, then it would normally be more rational to prefer the less tested drug. Similarly, if a difference exists between D_1 and D_2 with respect to the time it takes to bring about the cure, the costs involved, the inconvenience to the patients, etc., then all these factors may enter in and outweigh the lack of sufficient positive evidence for one of the drugs.

These considerations about usefulness may appear to undermine the whole idea that functionality and positive evidence are so central in medicine. But functionality is relevant to usefulness, because nothing is useful unless it works. The usefulness, or the utility, of a technique is a relative and subjective property. Usefulness depends on so many things and on the individual who will use it. What is useful to one person may not be useful to another, and what is useful in one context may not be useful in another. Also, unless the technique in reality gives rise to whatever effect the person who applies it thinks that it does, the person does not truly use that technology.

Suppose that Mr. Cough claims to have a therapy that can cure the common cold: the muffler. As soon as his throat begins to become sore, he puts the muffler around it, and after three or four days of coughing and sniveling, the cold is gone. It works every time, and Mr. Cough finds the therapy useful. But suppose now that a friend has persuaded him not to put the muffler on and see how his cold proceeds without it. The warmth around his throat is gone, but the cold disappears after three or four days just the same. What should we say about such a case? That the muffler method was useful as long as Mr. Cough thought that it was a functional cure, but that it was no longer useful when he discovered that it had no effect on the cold? Or that he had made a mistake about the usefulness of the muffler method all along, and that the mistake became obvious when he discovered that it did not work?

I would say that the second answer is the most reasonable. Functionality is a precondition for usefulness. The other condition is that the user subjectively values it as useful to him. Let me suggest the following definition:[1]

> T(E) is useful if, and only if, for some user X: (a) T(E) works sufficiently well, and (b) X views T(E) as useful.

By "sufficiently well," I mean that T(E) must function to a degree that is sufficient for being useful to the user. That degree may differ between different users and depend on their personal estimates of how efficient T(E) must be in order to be of use to them. But once this efficiency estimate is determined by the individual user, then the question of whether T(E) has that efficiency is objective. Therefore, (a) is an objective condition and (b) is a subjective one.

In the case of the muffler method, the efficiency did not concern what nature was doing all by itself. The user based his belief in the usefulness of the method on the false assumption that the muffler truly had an effect on the cold. Condition (b) was fulfilled, but he was wrong about condition (a), so the muffler method was not, in reality, useful.

Turning to the issue of quackery, and to pseudo-technology in general, it becomes plain how to treat the concept of usefulness. The user, Mr. Cough, for example, is the natural judge of whether a technique is useful to him. He is the authority on his problems, aims, and personal circumstances. He may still be mistaken, however, about whether the technique in question does work. The mistake is then due to a misconception about the true efficiency of the technique. A therapy that is falsely presented as a corroborated and working treatment is quackery. This is so even if the patient himself believes, and subjectively experiences, that it works and is useful to him.

3. Placebo

However, the problem is more complicated than this because of the existence of the so-called placebo effect. We have no real scientific explanation in physiological or biochemical terms for this phenomenon. Most medical therapies are associated with, and sometimes subtly interact with, those processes in the human mind that enhance the health and well-being of the patient. Some researchers have regarded the placebo effect as illusion and hoax. But as any practicing doctor knows, the way a medical treatment is administered often has some influence on how well it cures. From the patient's point of view, the most critical factor is that the illness is mitigated or cured, irrespective of how the cure was effected. If the medicine alone cures the patient, that is fine. If it was the medicine together with the patient's strong belief in the medicine that effected the cure, then that is fine too. The question is, should these psychological conditions be considered part of the therapy?

If we view such conditions as part of a therapy, then how can we say that homeopathy and magnetic belts are quackery? Both treatments probably have some positive effect on the health of the patients who use them. Yet as far as I know, no "real" effect of the therapies exists at all. We could argue that the only relevant element is the total effect, real or not. But we find it strange to claim that a therapy with no real effect "works"— especially when the reason it "works" is that the patient falsely believes that it does have a real effect. Homeopathy and magnetic belts have some effect because of the illusion that those therapies as such have an effect. In a way, they are parasitic on real effects. Once a patient becomes aware that those therapies possess no real effects, all effects are gone.

The other option is to try to make a careful distinction between the "therapy as such" and the way it is applied or used. I think we can do so in many cases, and then test the "real" functionality of the therapy in double-blind tests. However, I am not comfortable with the terminology here. If a placebo enhances a patient's health, nothing is unreal about that effect; a cure is a cure. We should rather talk about "pure therapy effects" (to be inferred from double-blind tests), "therapy-handling effects," and "pure handling effects." Each effect is equally real, although generated by different causes. Such a distinction would clarify the discussion about these issues.

The problem with homeopathy, magnetic belts, and the like is not that they have no basis in scientific theory. Many regular medical techniques do not either. Nor is the problem that none of them has any effect whatsoever, because many of them do. The problem is that all of these therapies are based on a false claim: that they are genuine and that a pure therapy effect exists. In my view, that is what makes them pseudo-therapies.

Genuine medical technology produces genuine medical techniques—it produces techniques that have a corroborated possibility of working. Quackery, however, is not genuine and does not produce anything but pure handling effects. Because no pseudo-therapies that work exist, they cannot be useful to anyone. This is so even if thousands of patients and doctors claim that they have great use of such therapies. Granted, patients undergoing these treatments may be helped or even cured, but they are not helped or cured by the pseudo-therapy; they are affected by the handling alone, a handling that is based on a faked therapy.

4. Science Again

Could the existence of scientific explanations be a good reason for claiming that a therapy is genuine? If a scientific theory tells us that too much of some substance in the body causes a sickness, and another theory tells us that some therapy will reduce that substance in the body, then we appear to have a good reason to recommend the therapy in question.

However, the problem is that we can always legitimately question the truth of the theories on which a therapy is based. If the theory happens to be true, then any therapy that is based on it will work if handled properly. But a scientific theory need not be true, only partly confirmed or corroborated. Many observations may have confirmed a scientific theory and the theory may still prove to be totally false on a basic, explanatory level. That false theories may have some true, observable consequences is the heart of the scientific problem of induction. Suppose that a physician in the 16th century recommends the therapy of bloodletting in accordance with the scientific theory of the day, Galenos' theory of the four bodily liquids. Why would the physician justify bloodletting as a medical treatment to combat the flu just because it is in accord with Galenos' theories? Without any independent corroboration of bloodletting as a helpful therapy in this situation, the theory alone is not a good argument for recommending it.

I have argued that whether a therapy is quackery does not depend on the existence of scientific explanations. Some genuine therapies have a scientific basis, while others do not. Likewise, we could argue that the quackery status of a therapy does not depend on the truth or falsity of the theories on which it is based. The central issue is whether it has tested successfully and whether it works. We are helped if we have a corroborated scientific theory that supports the therapy used, because such a theory gives us some explanation of how the therapy could work. But when considering the possibility of the therapy working, I believe that the existence of explanations is irrelevant.

We should also note that from a pragmatic—or technological—point of view, falsified theories are sometimes preferred to corroborated ones. For example, Newtonian mechanics is usually preferred to the Theory of Relativity when constructing cars and airplanes, because these theories overlap with respect to some observable, confirmed consequences (low speeds, weak gravitational fields, etc.), and Newtonian mechanics is simpler to use. From a pragmatic point of view, it is critical that researchers have confirmed the predictions of a theory within the range of practical interest. Whether the theories have proven false at other places is of less importance. An approximately true theory, like Newtonian mechanics, will suffice just as well as (or even better than) a non-falsified one as long as it is "true" within its range of application.

My claim, then, is that the empirical support of scientific theories (within a given range) does most of the work in "justifying" therapies. We should look for positive evidence that corroborates the theory and see if it also corroborates the possibility that the therapy works. What makes scientifically based therapies a rational choice is not the truth, the explanatory value, or the imbedding capability of scientific theories. This explains why even falsified theories may be helpful in constructing working techniques, and why a tested therapy, such as the use of acupuncture to mitigate pain, may be a rational choice even though it is not embedded in a scientific structure. It also explains why a therapy such as bloodletting was a bad idea, even though it was part of a once-established scientific theory.

I am aware that a few philosophers of science have discussed the role of science in relation to technological action. See, for instance, the discussion between Karl R. Popper (1974)[2] and Wesley C. Salmon (1981)[3], and for a fuller account, David Miller (1994)[4]. The problem with that discussion, as I see it, is that it has focused too much on empirical evidence for or against scientific theories and too little on the relation between empirical evidence and technology. David Miller and Karl Popper suggest that the function of scientific theory is to criticize alternative technologies. This suggestion, however, does not explain the positive role of false theories in technology, nor does it explain the rationality of choosing the corroborated alternative where this evidence is all we have.

The role of science in technology is much more complex than the often-used notion of "applied science" suggests (for an elaboration of this point, see Nordin, 1999)[5]; the point appears to be overlooked in the debate. As I see it, science is a heuristic and practical tool in technological research, but when we turn to the question of making a rational choice of technique, the explanatory power and coherence of scientific theories add nothing. Galenos' theory had both explanatory power and coherence in its time, and the compatibility or non-compatibility of a therapy to this theory would not be a rational guide in a practical choice situation. However, if the empirical evidence speaking for or against the theory also has direct implications for the functionality of the therapy in question, it would be relevant.

5. Alternative Medicine

Finally, I will try to link my analysis to what is usually called "alternative medicine." The terminology here is reasonably confused, and several names are in use to describe the concept. The European Commission launched the term "unconventional medicine," embracing a vast and, qualitatively speaking, heterogeneous group of therapies. It includes a large number of individual diagnostic and therapeutic procedures, as well as systematic and comprehensive concepts of health and disease. Depending on differing semantic nuances, unconventional medicine is also defined as "complementary medicine," "alternative medicine," "practical medicine," "natural medicine," "biological medicine," and "holistic medicine." The common denominator of the labels is their exclusion from the realm of conventional established scientific medicine and their sparse representation in research and teaching at universities. The commission has also said, "unconventional medicine often lacks a comprehensive evaluation of its efficacy (for historical reasons this applies to some conventional medical procedures as well)."[6] The World Health Organization sometimes uses the label TCAM, which stands for traditional, complementary, and alternative medicine.

I will not indulge in a semantic analysis of the different terms here. However, most terms are defined relative to social norms and conventions. What is "alternative" or not is relative to public health-care systems, professions, and training, and the use of the term may therefore change from culture to culture and from time to time. But alternative medicine also has a question mark attached to it: its unscientific basis and/or its lack of systematic evaluation. To put it bluntly, is alternative medicine just another name for quackery? No doubt, many professionals equate alternative medicine and quackery. But a growing awareness exists among both medical professionals and authorities that this is too simplistic. The uses of terms like "complementary medicine" and even "integrated medicine" show that conventional medicine is not altogether hostile to everything "alternative."

I have been talking about "genuine" medical therapies in contrast to pseudo-therapies. Quackery is a practice that uses pseudo-therapies as if they were genuine and working. Quackery is based on a falsehood. But since alternative and conventional medicine are defined on social conventions, no logical connection exists between alternative medicine and quackery. From a historical point of view, an empirical connection may well exist, and we may also count as an empirical fact that many alternative therapies are quackery. However, some alternative therapies both work and have been corroborated. They are "alternative" only in the sense that they happen to exist outside the established health-care system. For the same reason, we may ask whether pseudo-therapies exist within conventional medicine. If that is so, then such uses of pseudo-therapies are obviously not part of good, "true" medicine—but they may still be part of conventional medicine, established and legitimized by authorities and professionals.

6. Conclusions

I have taken the term "quackery" to be negatively value-laden. For that reason, the term is not suited to be defined for purely social/legal purposes in terms of profession and training. Not every therapeutic activity by non-professionals outside the established health-care systems is bad or fraudulent. For the same reason, it is not a good idea to define quackery as therapeutic activity that is not based on scientific theories. Many good, successful medical therapies—both inside and outside conventional medicine— are lacking in strict scientific explanation. I have suggested instead that quackery should be linked to pseudo-therapy, to false claims of the functionality of therapies.

Such a definition makes quackery independent of social conventions. As a consequence, we cannot equate alternative medicine with quackery. The use of terms like "complementary" or "integrative" medicine signals dissociation from such identification. Yet quackery exists in both conventional and alternative medicine. Wherever it occurs, we should consider it as a dubious enterprise. The negative effect of some quackery is obvious. Some pseudo-therapies are quite dangerous in themselves and may cause more harm than cure to the patient, while other uses of quackery prevent the patient from getting adequate, helpful treatment.

The paradoxical nature of quackery arises from the existence of placebos. Pseudo-therapies do not have a pure therapy effect and are not, therefore, useful to anyone. But the quackery may have positive effects on the patient's health anyway, due to its pure handling effect. In a situation where a doctor has no clear view of what is wrong with the patient or what to do with him, the doctor may be tempted to prescribe a placebo and tell the patient that it is a good, well tested cure. Doing so might even be rational in some sense. But quackery is parasitic on the trust that patients put in medical therapies in general. The moral dilemma is that with too much quackery, the good effects could be gone altogether.

NOTES

1. Ingemar Nordin, "The Rationality of Technology," *Science Studies*, 2, (1989), pp. 4–5.
2. Karl R. Popper, "Reply to my Critics," in P. A. Schilpp, *The Philosophy of Karl Popper*, (La Salle, Illinois: Open Court, 1974), p. 1025.
3. Wesley C. Salmon, "Rational Prediction," *British Journal of The Philosophy of Science*, 32, (1981), pp. 115–125.
4. David Miller, *Critical Rationalism: A Restatement and Defence*, (Chicago: Open Court, 1994), pp. 20f.
5. Nordin, "The Role of Science in Medicine," *Theoretical Medicine and Bioethics*, 20, (1999), pp. 229–244.
6. J. Monckton, B. Belicza, W. Betz, H. Engelhart, and M. Van Wassenhoven, *COST Action B4*, European Commission, European Communities, (1999).

Thirteen

PRIVATE LIFE: SKETCHING ENTRANCES FOR ETHICAL ANALYSIS

Nina Nikku

1. Introduction

We may all face situations where we apprehend that what we consider as intimate, private zones of our life have been exposed to intrusion. This may happen when a person becomes dependent on others for assistance, as in a case of disease or impairment. Examination, diagnosis, and treatment may involve physical intrusion and may force a disclosure regarding the patient's habits and behavior. Living at a nursing home or being dependent on receiving help in our home may lead to a radical change of the limits of privacy. Health-promotion, disease prevention, and working for health often involve actions that aim at influencing the care recipient's lifestyle. Giving advice on health-promoting behavior, sharing information about health risks, and encouraging healthy habits are all activities with a beneficent aim, but we may apprehend them as intrusions into private matters. Regulations, laws, and political steering are measures that may interfere even more in the individual's private sphere.

Everyone will probably agree that some parts of our lives are, and ought to be, private. However, our apprehension of what is private may differ depending on factors such as cultural context or our relation to the intruding person. Our opinions on what may be considered as an ethically acceptable intrusion into the private sphere and what is not may vary. Circumstances like the individual's condition or the way in which the intrusion is performed may affect our attitudes on the topic. The starting-point for this essay is that the limits of private life are not a given. The essay aims to illuminate the multidimensional nature of the notion private life. By giving attention to the complexity of demarcations in private life, we may develop an ethical sensitivity and consciousness in these situations. This task is urgent, especially in relation to health care practice, as therapeutic encounters have in many ways moved from the clinic to the client's private home. This work may in turn have consequences for how we should perform ethical analysis, as such an analysis must consider the limits of private life.

This text begins by problematizing the content of the notion of private life. I will exemplify different entrances to analyzing private life and demonstrate how to place private life in the context of society and public life, arguing that doing so will give us a richer understanding of the limits and extent of the notion. Thereby I will suggest an alternative methodology that uses an open form for analysis in its ethical scrutiny and considers a cross-cultural perspective.

2. Private Life and Privacy

The notion "private" is defined in dictionaries as not public or official, but personal, internal, not to be revealed to others, and secret. What is private belongs to or exists for the use of one particular person. The concept is described as the condition of being undisturbed, alone, or free from interference. The opposite of private is public, a term that is described as open, concerning people in general, and available for everyone. While western societies consider it valuable that some matters are public and openly accounted for, the respect for an individual's private life is fundamental. Defending the importance of respect for what we consider as belonging to the private part of our lives is often based on regard for human dignity. The ethical norm is referred to as "the rule of privacy," used to convey the condition of having a private life that is free from interference. Consequently, an analysis about whether privacy is respected requires knowledge of the content and limits of private life.

Discussions of the rule of privacy often focus on limited or restricted access to information about a person. This aspect of the rule aims at the moral imperative not to investigate or spread information about a person without his or her permission. In this reading—a narrow interpretation—the rule is connected to confidentiality. In a broader understanding, the ethical norm focuses also on limited access to a person, including limitations on intervening in the zones of intimacy, secrecy, anonymity, and solitude (Beauchamp and Childress, 1989).

Some general opinions of what may be a violation of someone's private life exist, but what we strictly consider as an infringement or what the person himself may apprehend as a sense of loss of privacy may vary. Since the limits of private life are socially and culturally created and since societies are different, the need for privacy will historically differ from one society to another—as well as among different groups in one society. No particular item may be intrinsically private. Charles Fried's example, which illustrates different aspects of an individual's privacy in relation to other persons (Fried, 1984), describes well the way in which an infringement of the private life is complicated in general descriptions:

> We may not mind that a person knows a general fact about us, and yet feel our privacy invaded if he knows the details. For instance, a causal acquaintance may comfortably know that I am sick but it would violate my privacy if he knew the nature of the illness. Or a good friend may know what particular illness I am suffering from, but it would violate my privacy if he were actually to witness my suffering from some symptom which he must know is associated with the disease.

The example shows that the limits of private life may vary depending on our relations to other persons. The same individual may have several gradations of his private life; he may consider different parts as private, more private, and most private. The individual's relations to other persons may be included in those gradations. One grade of private life may include the closest family; another may cover the relations to friends and relatives, etc. Consequently, whether an action may be an infringement into private life is due, among other things, to whether the actor is part of the individual's private life.

The experience of Jean L. Briggs, an anthropologist who studied a remote Eskimo community, illustrates that a broad cultural difference exists in practice and in our apprehension of private life. Briggs experienced an invasion of her private life during her stay with the tribe. Her tent was never empty from the time she awoke in the morning (sometimes sooner) until she retreated to her sleeping bag in the evening. Later she learned that it was not Eskimo etiquette that imposed the obligation of constant attendance on a visitor's wants, but a social obligation. Where people live in a potentially dangerous environment, we find a strong emphasis on reciprocity and responsiveness to the needs of others (Briggs, 1970; see also Moore, 1984). The Eskimos did not conceive of what felt to Briggs like a trying intrusion into her private life in terms of restrictions of privacy. However, the cultural difference between Briggs and the Eskimos may be apprehended in several ways:

(1) The anthropologist Briggs with her western background assumed that her tent was a private area, but the Eskimos did not apprehend the tent in this way.
(2) The Eskimos considered the tent as an area for their private life, but due to circumstances, intrusions were accepted and therefore not conceived as any wrongful violation of privacy.
(3) The tent is considered as private, and the intrusion of others is an infringement of a prima facie right to privacy; however, the interest of the collective overrides the right to privacy.

Briggs concluded that the Eskimos' construction of private life resulted from the harsh environment in which their lives take place. A tent belonged to those living there, but was not understood as private in the sense of implying restrictions. An example of this point was an unsuccessful attempt to introduce the custom of knocking on the door before entering a tent.

3. Developing Ethical Tools

No clear and simple limits exist as to what we might apprehend as belonging to the private section of our life. The notion of private life remains too complicated to be described in one dimension. Using models that illuminate different perspectives, we can increase our possibility of understanding the demarcations of private life in settings like distinct cultures or subcultures and develop receptiveness for situations where an intrusion may be ethically sensitive. A penetrating analysis of the ethical judgment of infringements of individuals' private lives may result from this method.

I will sketch three such ways of illustrating the numerous dimensions to what we may reckon as the individual's private life. The approaches may be used to describe the private in relation to the public, a private domain, and an intimate zone. The three points of entry are each applicable to different questions. The first one focuses on the individual's private life in relation to public society; the second is useful in an interpretation of the ethical norm of informed consent in relation to private life; and in the third approach, the question of the individual's intimate core comes to the fore. For a richer analysis of a predicament concerning the private life, we may use more than one approach.

4. Private Life in Context

I will begin by mapping out a way to place private life in context. In a discussion about private life and the difficulties in establishing this notion, grasping the framework of public and private life in a wide perspective—thereby including the possibility of illustrating changes in the correlation—is necessary. The sketch below demonstrates the relationship between private and public, and represents a way of understanding how the private interacts with other components of human life. It also announces that outside circumstances may cause us to view the individual and what is often considered as belonging to his or her private life in another way. The arrows between the categories indicate the alteration that may take place in a given case over time, in comparison to different cultural contexts, or as a consequence of other circumstances. The different components of the scheme are presented briefly below.

Beyond Individuality—Society

↑↓

The Individual's Public Life

↑↓

The Private Domain

↑↓

The Intimate Zone

↑↓

Beyond Individuality—Organism

Recall Briggs, the anthropologist who went to a remote Eskimo tribe and never felt that she had any privacy: In such a harsh environment, the individual and his or her individuality may have to stand back. This example of a tightly knit community illustrates the first classification in the model above. Under those circumstances, we are not chiefly recognized as individuals but as parts of society; here we are a part of a group, a collective, or a stream steered by generally applicable regulations and laws. The individual is subordinated to societal goals.

The next category concerns the individual's public life. We show this part of our lives and our selves without apprehension. Performing as individuals in social settings, interacting with others at work, and engaging in leisure time are activities of public life. Most of a person's daily activities can be regarded as public. Walking in the street, driving my car, and having dinner at a restaurant are all examples of every day occurrences categorized as public life.

The content of private life that is viewed as a private domain and an intimate zone is further described below. In these categories we find the main concern of this essay: the core of what we usually define as private life. A tension exists between public life and private life, but the demarcation is transparent. Cultural differences may cause problems in situations involving different generations, ethnic groups, and social categories.

The section beyond the intimate zone (and beyond individuality) occurs when the individual is scrutinized in a completely organism-centered perspective. Here we view the biological creature, not the separate person. The unconscious person in a heart-lung machine is such an organism if the limits of the private for that person are formulated in a different way than they are for a conscious and reflecting individual. This must not mean that the person is treated in a disrespectful way, but that the perspective is not focused on the personal. Instead, the focus is on the body machinery or the human being, on the level of cells or genes. Consider the example of organ transplantation. In an operation, the innermost part of the patients is revealed: intimate parts of the body, feelings like fear and anguish, and traces of previous living. However, the concern and perspective of the operation team is neither the private life nor the intimate sphere of the patients as individuals. Their main concern is the human machinery, its organs, and their function. This perspective does not prevent the surgeon from being a compassionate fellow being before as well as after the operation.

The sketch illustrates how a broad perspective may help us to describe and further analyze private life in connection to its context. The scheme may, for example, clarify differences between private life and public life in diverse cultural contexts. I will further clarify the perspective of private life (under the private domain and intimate zone sections) below.

5. Private Life Viewed as a Domain

Another entrance into the concept of private life lies in considering it as a domain. This provides an approach for understanding the content of private life that draws attention to the different areas of private life. Actions directed toward the individual and her closest surroundings might be regarded *prima facie* as intrusions. This domain includes the following elements:

(1) the individual's body and products from the body, for example, blood or tissue;
(2) the individual's intimate relationships;
(3) the individual's private products, such as writings or drawings;
(4) the individual's home and belongings like clothes.

Different actions directed against this domain—like being touched, observed, listened to, or investigated—may be regarded as potentially intrusive into the private life.

Using the model of a domain highlights an area that embraces the individual and her closest surroundings. We should not regard actions directed toward the private domain as ethically unacceptable violations. In several cases, the individual herself has asked for or given her permission to these actions. This is true, for example, of most examinations and tests that are carried out in the health care field, where the rule of informed consent is a basic ethical norm. However, we cannot regard informed consent to intrusions into the private domain in the same way as the rather hard claim for treatment. By entering the public arena, the individual gives her indirect or tacit consent for being observed in the sense of being watched by other people. She cannot ignore that she will be heard when speaking. In these cases, we cannot talk about an unacceptable intrusion into the individual's private life. The rule of informed consent is not applicable, or else must be interpreted in a different way than it currently is. The criteria of proper informed consent are that the person (1) is informed; (2) understands the information; (3) is competent; (4) is not exposed to pressure; and (5) gives voluntary and active consent (Faden and Beauchamp, 1986).

Privacy, normally considered as a value worthy of respect, is sometimes seen in terms of a right. The right to privacy can be regarded as the right to self-determination, applied to the private domain. The right to privacy is, then, a part of the right to self-determination, but is more extensive because the right to self-determination may be restricted with reference to the individual's condition (Nikku, 1997). A non-autonomous person is not deemed to have the corresponding right to self-determination. But consider the example of an unconscious person who is undressed and examined in a corridor at the hospital where people are passing through all the time. Even if this person is not aware of this humiliating situation, we should consider the action of undressing and examining him in front of all the people passing as an intrusion upon his privacy.

While the right to self-determination is restricted with reference to the individual's capability to understand and decide a pertinent question, a right to privacy is not connected to the requirement of competence. Instead another criterion must exist for how to analyze a potentially unethical intrusion when the norm of informed consent is not applicable. The example of persons providing daily assistance to the disabled, whom we will call personal assistants, shows alternative ways of dealing with an ethically sensitive situation.

Personal assistants daily have to handle the ethical dilemma of intrusions into a private domain. A study of the occupation's applied ethics shows that these workers develop different strategies for managing difficulties (Nikku, 2002). The client has invited the personal assistant into his or her private life to assist in the activities of daily life. The client has approved the entrance into the domain, but no specified agreement exists on what tasks will be called for in this domain. The assistant can perform as "the client's hands and feet," strictly doing what he or she is told. Some clients prefer this model. However, several clients favor at least some initiative on the part of assistants concerning their work, but this may make the ethical norm concerning consent inapplicable. In such cases, the assistants in the study cited above employed tactics in their work such as settled routines, continuing communication about their actions, or sensitive questions:

(1) Settled routines—the assistant works according to a scheme in which the tasks are decided beforehand and preferably written;
(2) Continuing communication—the assistant is continuously communicating what he or she is going to do, thereby giving the client opportunity to disapprove;

(3) Sensitive questions—the assistant uses small-talk questions and sensitivity, listening for the client's approval.

These tactics were used as alternatives to formal informed consent, but still represented respect for the disabled client's private domain.

6. The Intimate Zone

I will turn to the second section of private life in the model sketched above: the intimate zone. Intimacy is often connected to intimate parts of our body or to activities such as sexual behavior. This may not always be true. Consider the prostituted woman who keeps parts of her body off-limits to her customers or forbids some actions. Some prostitutes will not allow customers to touch their hair or kiss them during the sexual act. Through these exceptions, she avoids sharing her intimacy with the partner.

In an ideal state, the intimate zone—as well as shared intimacy—is accessible only for some select persons. Other people may violate our intimate zone and intimacy, on purpose or unintentionally. Even if the boundaries of the intimate zone vary from person to person, factors in our lives that have highly charged emotions attached to them tend to be most intimate. A teenager's diary about love and future dreams is an example of recording such elements for ourselves.

In the study of applied ethics mentioned above, a woman providing assistance for disabled persons regarded an action like questioning a client's judgment as potentially a worse violation of that person's intimacy than helping him to the toilet. Receiving help and assistance in your home is an intrusion into the private life. If the need also includes intimate care with personal matters, a tension may occur in the relation between the assistant and the assisted person. The personal assistants in that study believed that we can handle even intimate situations with respect. Interpreting the assistants' work in connection to the model, we may say that these actions are performed in the intimate zone, but if the assistant is able to handle the situation in a professional way, no sense of uneasiness or embarrassment will occur. Acting in a professional way in such a case means that the charged intimate realm is decoded into what we perceive as the ordinary; the intimacy of the client is not intruded.

The ability to transform uneasy situations and incursions into the intimate zone into something ordinary or neutral is a professional quality. This skill, as well as the ability to switch between different roles and functions, is required of an ideal health care professional. Take the example of people who provide daily assistance to the disabled, for whom one desired role is to be the client's hands and feet. Acting in such an instrumental way is an example of how to handle an intimate situation and transform it into an ordinary one. However, several clients prefer some initiative from the assistants concerning their work, and this involves taking a different position in other situations.

Recall the personal assistant who described the act of questioning a client's judgment as an example of violating that person's intimacy. This transgression may consist of interfering into personal matters like arranging the home, or taking too much command over the disabled person's daily life. In a case where the client apprehends that the assistant has intruded upon the expression of his innermost self and his personality, this is not only an infringement of the person's private domain, but also of his intimacy. This may not be ethically acceptable.

However, the work of some professions intrudes not only into the bodily intimate zone but also into the intimate mental context of their clients. Methods like psychoanalysis and hypnosis aim at interfering in the patient's most intimate zone. In those fields the ethical norm of informed consent cannot be appropriate, since the patient may not know what will be revealed. Such a relation must rely on an unconditional trust and absolute confidentiality; a professional in that type of situation approaches a shared intimacy.

7. An Open Form of Ethical Analysis

Using a broad model for analysis provides possibilities for understanding the limits of private life in different social and cultural settings without the restrictions of a formal definition of what private life is. Choosing this open structure for ethical analysis can be an alternative to traditional forms, as the demand of universality that attends formal definitions tends to make applied ethics less applicable. The examples above show that the limits of private life may differ radically in time and space, depending on individual circumstances and other factors. Independent of these difficulties, finding ways for an ethical analysis of the limits of private life and of the criteria for ethically acceptable infringements is urgent. A need exists for a theoretical basis for a further developed ethics of private life. The need is perhaps most critical in questions related to health care matters, as the limits of private life today have a new, deeper role in therapeutic and nursing encounters, one that traditional biomedical ethics did not face. This essay intends to point out that

WORKS CITED

Antonovsky, Aaron. (1982) *Health, Stress, and Coping.* London: Jossey-Bass Publishers.

———. (1987) *The Mystery of Health.* London: Jossey-Bass Publishers.

Barry, B. (1965) *Political Argument.* London: Routledge & Kegan Paul.

Beauchamp, Tom L. and James F. Childress. (2001) *Principles of Biomedical Ethics*, 5th edit.. Oxford: Oxford University Press.

Bentham, Jeremy. (1996) *An Introduction to the Principles of Morals and Legislation*, ed. J. H. Burns and H. L. A. Hart. Oxford: Clarendon Press.

Bergin, Allen E., and Sol L. Garfield, eds. (1994) *Handbook of Psychotherapy and Behaviour Change.* New York: John Wiley & Sons.

Binswanger, Ludwig. (1958) "The Case of Ellen West." In *Existence: A New Dimension in Psychology and Psychiatry*, eds. R. May, E. Angel, and H. Ellenberger. New York: Basic Books, pp. 237–94.

Blum, Harold P. (1992) "Psychic Change: The Analytic Relationship(s) and Agents of Change," *International Journal of Psycho-Analysis*, 73, pp. 255–265.

Boorse, Christopher. (1975) "On the Distinction Between Disease and Illness," *Philosophy and Public Affairs*, 5, pp. 49–68.

———. (1976) "What a Theory of Mental Health Should Be," *Journal for the Theory of Social Behaviour*, 6, pp. 61–84.

———. (1976) "Wright on Functions," *Philosophical Review*, 85, pp. 70–86.

———. (1977) "Health as a Theoretical Concept," *Philosophy of Science*, 44, 542–573.

———. (1997) "A Rebuttal on Health." In *What is Disease? Biomedical Ethics Reviews*, eds. J. Humber and R. Almeder. Totowa, N. J.: Humana Press.

Boss, Medard. (1979) *Existential Foundations of Medicine and Psychology.* New York: Jason Aronson.

Bourne, Peter G., Robert M. Rose, and John W. Mason. (1968) "17-OHCS levels in combat Special Forces 'A' Team under threat of attack," *Archives of General Psychiatry*, 19, pp.135–140.

Bowlby, John. (1973) *Attachment and Loss*, vol. I: *Attachment.* New York: Basic Books.

Brandtstadter, Jochen, Bernhard Baltes-Götz, Clemens Kirschbaum, and Dirk Hellhammer. (1991) "Developmental and personality correlates of adrenocortical activity as indexed by salivary cortisol: Observations in the age range of 35 to 65 years," *Journal of Psychosomatic Research*, 35, pp.173–185.

Briggs, Jean L. (1970) *Never in Anger: Portrait of an Eskimo Family.* Cambridge, Mass: Harvard University Press.

Brülde, Bengt. (1998) "Vad är Hälsa? Några Reflektioner Kring Hälsobegreppet" [What is Health? Some Reflections on the Health Concept], Göteborg, Sweden: Department of Philosophy, University of Gothenburg.

———. (2000) "On How to Define the Concept of Health: A Loose Comparative Approach," *Medicine, Health Care and Philosophy*, 3, pp. 305-308.

Buchanan, Allen. (1989) "Health-Care Delivery and Resource Allocation." In *Medical Ethics*, ed. Robert M. Veatch. Sudbury, Mass.: Jones and Bartlett Publishers, pp. 321–361.

Bullington, Jennifer. (1999) *The Mysterious Life of the Body: A New Look at Psychosomatics*. Linköping, Sweden: Linköping Studies in Arts and Science, No. 190.

Bury, Michael. (1982) "Chronic Illness as Biographical Disruption." *Sociology of Health and Illness*, 4, 167–82.

———. (1988) "Meanings at Risk: The Experience of Arthritis." In *Living with Chronic Illness*, eds. Robert Anderson and Michael Bury. London: Unwing Hyman, pp.89-116.

———. (1991) "The Sociology of Chronic Illness: A Review of Research and Prospects," *Sociology of Health and Illness*, 4, pp. 451- 468.

———. (1997) *Health and Illness in a Changing Society*. London: Routledge & Kegan Paul.

Byrne, P. S, and B. E. L. Long. (1984) *Doctors Talking to Patients*. Exeter: The Royal College of General Practitioners.

Canguilhem, Georges. (1978) *On the Normal and the Pathological*. Dordrecht: D. Reidel Publishing Company.

Cannon, Walter. (1935) "The Stresses and Strains of Homeostasis," *American Journal of Medical Science*, 189, pp.1–14.

Charmaz, Kathy. (1983) "Loss of Self: A Fundamental Form of Suffering in the Chronically Ill," *Sociology of Health and Illness*, 5, pp.168–195.

———. (1991) *Good Days, Bad Days*. New Brunswick, N. J.: Rutgers University Press.

Coderre, T. J., J.Katz, A.L. Vaccarino, and R. Melzack. (1993) "Contribution of Central Neuroplasticity to Pathological Pain: Review of Clinical and Experimental Evidence," *Pain*, 52, pp. 259–285.

Conrad, Peter. (1990) "Qualitative Research on Chronic Illness," *Social Science and Medicine*, 30, pp. 1257–1263.

Corbin, Juliet M. and Anselm Strauss. (1988) *Unending Work and Care*. San Francisco: Jossey-Bass Publishers.

Cumming, Elaine and W. E. Henry. (1961) *Growing Old*. New York: Arno Press.

Danziger, K. (1997) *Naming the Mind: How Psychology Found its Language*. London: Sage Publications.

DSM-IV. (1994) *Diagnostic and Statistical Manual of Mental Disorders IV*. Washington D. C.: American Psychiatric Association.

Elliott, Robert and Elisabeth James. (1989) "Varieties of Client Experience in Psychotherapy: An Analysis of the Literature," *Clinical Psychology Review*, 9, pp. 443–467.

———. and D.A. Shapiro. (1992) "Client and Therapist as Analysts of Significant Events." In *Psychotherapy Process Research: Paradigmatic and Narrative Approaches*, eds. G. Shaké, G. Toukmanian, and David L. Rennie. Newbury Park: Sage Publications.

Erikson, Erik Homburger. (1982) *The Life Cycle Completed*. New York: Norton.

Faden, Ruth R. and Tom L. Beauchamp. (1986) *A History and Theory of Informed Consent*. Oxford: Oxford University Press.

Falk, Hans, L. Gustavsson, and B. Richt. (2001) *Kontaktmannaskap: En väg till bättre äldreomsorg* [Contact Persons: A Road to Improved Elderly Care]. Stockholm, Sweden: Liber.

Fox Keller, Ellen. (1992) "Nature, Nurture and the Human Genome Project." In *The Code of Codes, Scientific and Social Issues in the Human Genome Project*, eds. D. J. Kevles and L. Hood. Cambridge, Mass.: Harvard University Press.

Frank, Arthur W. (1995) *The Wounded Storyteller*. Chicago: University of Chicago Press.

Frankl, Viktor E. (1966) "What Is Meant by Meaning?" *Journal of Existentialism*, 7, pp 21–28.

Freud, Sigmund. (1905/1960) "On Psychotherapy." *The Standard Edition of the Complete Psychological Works of Sigmund Freud. Vol. VIII*. London: The Hogarth Press.

Fried, Charles. (1984) "Privacy: A Rational Context." In *Philosophical Dimensions of Privacy*, ed. Ferdinand David Schoeman. New York: Cambridge University Press, pp. 203–22.

Friedhoff, Arnold J. and Jeannette C. Miller. (1983) "Clinical Implications of Receptor Sensitivity Modification," *Annual Review of Neuroscience*, 6, pp. 121–148.

Fulford, K. W. M. (1989) *Moral Theory and Medical Practice*. Cambridge, England: Cambridge University Press.

Gadamer, Hans-Georg. (1993) *Über die Verborgenheit der Gesundheit*. Frankfurt am Main: Suhrkamp Verlag.

Gould, Stephen J. and E. S. Vrba. (1998) "Exaptation: A Missing Term in the Science of Form." In *Nature's Purposes: Analyses of Function and Design in Biology*, eds. C. Allen, M. Bekoff, and G. Lauder. Cambridge, Mass.: The MIT Press.

———. (2000) "More Things in Heaven and Earth." In *Alas, Poor Darwin: Arguments Against Evolutionary Psychology*, eds. Hilary Rose and Steven Rose. London: Jonathan Cape.

Grawe, Klaus and Mariann Grawe-Gerber. (1999) "Ressourcenaktivierung: Ein Primäres Wirkprinzip der Psychotherapie," *Psychotherapeut*, 44, pp. 63–73.

Gullacksen, Ann-Christine. (1998) *När smärtan blir en del av livet* [When Pain Becomes a Part of Life]. Lund, Sweden: University of Lund.

Häring, Bernard. (1972) *Medical Ethics*. Slough: St Paul Publications.

Harris, A. J. (1999) "Cortical Origin of Pathological Pain," *Lancet*, 354, pp. 1464–1466.

Heidegger, Martin. (1986) *Sein und Zeit*, 16th edit.. Tübingen: Max Niemeyer Verlag.

———. (1994) *Zollikoner Seminare*. Frankfurt am Main: V. Klostermann.

Hellström, O. (1993) "The Importance of a Holistic Concept of Health for Health Care. Examples From the Clinic," *Theoretical Medicine*, 14, pp. 325–342.

———. (1998) "Dialogue Medicine: A Health-Liberating Attitude in General Practice," *Patient Education and Counseling*, 35, pp. 221–231.

———. (1999) *Patienten som person. Om mening och dialog i allmänmedicinsk praktik*
[The Patient as Person. On Meaning and Dialogue in General Practice]. Umeå, Sweden: Umeå University Medical Dissertations. New Series, No 585.

———. and J. Bullington, G. Karlsson, P. Lindqvist, and B. Mattsson. (1999) "A Phenomenological Study of Fibromyalgia: Patient Perspectives," *Scandinavian Journal of Primary Health Care,* 17, pp. 11–16.

Horvath, Adam O. (2000) "The Therapeutic Relationship: From Transference to Alliance," *Journal of Clinical Psychology,* 56, pp. 163–173.

House, James S., Cynthia Robbins, and Helen L. Metzner. (1982) "The Association of Social Relationships and Activities with Mortality: Prospective Evidence from the Tecumseh Community Health Study," *American Journal of Epidemiology,* 116, pp. 123–140.

Jahoda, Marie. (1958) *Current Concepts of Positive Mental Health.* New York: Basic Books.

Jakobsson, Einar. (1994) *Psykoterapins Uppgift: Hälsa, Bot och Självförbättring i Modernt Psykoanalytiskt Tänkande* [The Purpose of Psychotherapy: Health, Cure and Selfimprovement in Modern Psychoanalytic Thinking]. Linköping, Sweden: Linköping Studies in Arts and Science, No 108.

———. (2000) "How Specific Can the Psychotherapeutic Intervention Be?" Paper presented at The 4th International Conference on Philosophy and Psychiatry: "Madness, Science, and Society," Florence, Italy, 27 August, 2000.

Jung, Carl Gustav. (1984) *Modern Man in Search of a Soul.* London: Ark.

Kaplan, George A., Jukka T. Salonen, Richard D. Cohen, Richard J. Brand, Leonard Syme, and Pekka Puska. (1988) "Social Connections and Mortality from all Causes and Cardiovascular Disease: Prospective Evidence from Eastern Finland," *American Journal of Epidemiology,* 128, pp. 370–380.

Kelly, Michael, P. (1992) *Colitis.* London: Routledge & Kegan Paul.

Kernberg, Otto F. (1975) *Borderline Conditions and Pathological Narcissism.* New York: Jason Aronson, Inc.

Khushf, George. (2001) "What is at Issue in the Debate about Concepts of Health and Disease? Framing the Problem of Demarcation for a Post-Positivist Era of Medicine." In *Health, Science, and Ordinary Language*, ed. Lennart Nordenfelt. Amsterdam: Rodopi, pp. 123-169.

Kirschbaum, Clemens, Thomas Klauer, Sigrun-Heide Filipp, and Kirk Helmut Hellhammer. (1995) "Sex-specific Effects of Social Support on Cortisol and Subjective Responses to Acute Psychological Stress," *Psychosomatic Medicine,* 57, pp. 23–31.

Klein, Melanie. (1960/1975) "On Mental Health." In *Envy and Gratitude and Other Works 1946–1963.* London: The Hogarth Press.

Kleinman, Arthur. (1988) *The Illness Narratives.* New York: Basic Books Inc.

Koch, Lene. (1999) "Predicitive Genetic Medicine — A New Concept of Disease." In *Genetics in Human Reproduction,* eds. Elisabeth Hildt, and Sigrid Graumann. Aldershot: Ashgate.

Kohut, Heinz. (1984) *How Does Analysis Cure?* Chicago: The University of Chicago Press.

Kushner, Thomasine. (1981) "Doctor-Patient Relationship in General Practice a Different Model," *Journal of Medical Ethics,* 7, pp. 128–131.

Kutchins, Herb and Stuart A. Kirk. (1997) *Making Us Crazy. DSM: The Psychiatric Bible and the Creation of Mental Disorders.* New York: The Free Press.

Lag om åtgärder i forskning- eller behandlingssyfte med befruktade ägg från människa [Act concerning Measures for Purposes of Research or Treatment involving fertilized Human Ova]. (1991) SFS 1991: 115.

Liss, Per-Erik. (1993) *Health Care Need*. Aldershot: Avebury.

———. (1996) "On the Notion of a Goal: A Conceptual Platform for Setting of Goals in Medicine." In *The Goals and Limits of Medicine*, eds. Lennart Nordenfelt and Per-Anders Tengland. Stockholm, Sweden: Almqvist & Wiksell International, pp. 13–31.

Lloyd-Richards, Robert. (2000) "Death, Bereavement, and Spiritual Issues." In *Nursing In Haematological Oncology*, ed. Maggie Grungy. Balliere Tindall: ORT.

Luborsky, Lester and Paul Crits-Cristoph. (1998) *Understanding Transference: The Core Conflictual Relationship Theme*, 2nd edit.. Washington, D.C.: American Psychological Association.

Malterud, Kirsti. (1990) *Allmennpraktikerens möte med kvinnelige pasienter* [The General Practitioner's Meetings with Female Patients]. Oslo, Norway: Tano.

Maslow, Abraham H. (1954) *Motivation and Personality*. New York: Harper and Row.

May, Rollo, E. Angel, and H. Ellenberger, eds. (1958) *Existence: A New Dimension in Psychology and Psychiatry*. New York: Basic Books Inc.

———. (1983) *The Discovery of Being, Writings in Existential Psychology*. New York: Basic Books Inc.

Mayr, Ernst. (1982) *The Growth of Biological Thought: Diversity, Evolution, and Inheritance*. Cambridge, Mass.: Harvard University Press.

McDowell, Ian and Claire Newell. (1996) *Measuring Health: A Guide to Rating Scales and Questionnaires*, 2nd edit.. Oxford: Oxford University Press.

McEwen, Bruce S. and Elliot Stellar. (1993) "Stress and the Individual: Mechanisms Leading to Disease," *Archives of Internal Medicine*, 153, pp. 2093–2101.

———. (1998) "Protective and Damaging Effects of Stress Mediators," *The New England Journal of Medicine*, 338, pp. 171–179.

McWhinney, Ian R. (1997) *A Textbook on Family Medicine*, 2nd edit.. Oxford: Oxford University Press.

Meissner, W. W. (1991) *What is Effective in Psychoanalytic Therapy: The Move from Interpretation to Relation*. Northvale, New Jersey: Jason Aronson.

Mill, John Stuart. (1972) *Utilitarianism, On Liberty, and Considerations on Representative Government*. London: J. M. Dent & Sons Ltd.

Merleau-Ponty, Maurice. (1942/1963) *The Structure of Behavior*. Boston: Beacon Press.

———. (1945/1962). *Phenomenology of Perception*, London: Routledge & Kegan Paul.

Moore, Barrington Jr. (1984) *Privacy: Studies in Social and Cultural History*. New York: M. E. Sharp, Inc.

Nikku, Nina. (1997) *Informative Paternalism: Studies in the Ethics of Promoting and Predicting Health*. Linköping, Sweden: Linköping Studies in Arts and Science, No 155.

———. (2002) *Att bli sin brukares händer och fötter* [Becoming the Client's Hands and Feet] (Forthcoming in Swedish).

Nordenfelt, Lennart. (1987) *On the Nature of Health: An Action-Theoretic Approach*, Dordrecht: D. Reidel Publishing Company .
———. (1993a) *Quality of Life, Health and Happiness*. Aldershot: Avebury.
———. (1993b) "Concepts of Health and Their Consequences for Health Care," *Theoretical Medicine*, 14, pp. 277–285.
———. and Bo Petersson, and Per-Erik Liss. (1994) "Att Mäta Hälsa" [Measuring Health]. In *Livskvalitet* [Quality of Life] ed. Bengt Richt. Linköping, Sweden: Studies on Health and Society, 21, pp. 77–94.
———. (1995a) *On the Nature of Health: An Action-Theoretic Approach*, 2nd revised edition. Dordrecht: Kluwer Academic Publishers.
———. (1995b) "On Chronic Illness and Quality of Life: A Conceptual Framework," *Health Care Analysis*, 3, pp. 290–298.
———. (2000) *Action, Ability, and Health: Essays in the Philosophy of Action and Welfare*, Dordrecht: Kluwer Academic Publishers.
———. (2001) *Health, Science, and Ordinary Language*. Amsterdam: Rodopi.
Nozick, Robert. (1974) *Anarchy, State, and Utopia*. Oxford: Basil Blackwell Ltd.
Nussbaum, Martha C. (1994) *The Therapy of Desire*. Princeton: Princeton University Press.
Parsons, Talcott. (1955) *The Social System*. Glencoe, Ill., Free Press.
Plato. (1998) *Complete Works*, ed. John M. Cooper, Indianapolis: Hackett Publishing Company.
Priorities in Health Care: Ethics, Economy, Implementation. Final report by the Swedish Parliamentary Priorities Commission. Department of Social Affairs, SOU 1995:5. Stockholm: Fritzes.
Pörn, Ingmar. (1984) "An Equilibrium Model of Health." In *Health, Disease, and Causal Explanations in Medicine*, eds. Lennart Nordenfelt and B.I.B. Lindahl. Dordrecht: D. Reidel Company, pp. 3–9.
———. (1993) "Health and Adaptedness," *Theoretical Medicine*, 14, pp. 295–304.
Reisine, Terry, Hans-Urs Affolter, Geneviève Rougon, and Jacques Barbet. (1986) "New insights into the molecular mechanisms of stress," *Trends in Neuro Science*, 9, pp. 574–579.
Roth, Anthony and Peter Fonagy. (1996) *What Works for Whom? A Critical Review of Psychotherapy Research*. New York: The Guilford Press.
Saltman, Richard B. (1994) "Patient Choice and Patient Empowerment in Northern European Health Systems: A Conceptual Framework," *International Journal of Health Services*, 24, pp. 201–229.
Sandahl, Christer, Eva Birgerson, Bengt-Åke Armelius, Mats Fridell, and Vera Segraeus. (1998) *Bedömningsinstrument inom Behandling och Forskning* [Evaluation Instruments for Treatment and Research]. Stockholm, Sweden: SIS.
Sandler, Joseph and Anna Ursula Dreher. (1996) *What do Psychoanalysts Want? The Problem of Aims in Psychoanalytic Therapy*. London: Routledge & Kegan Paul
Schoeman, Ferdinand David. (1992) *Privacy and Social Freedom*. Cambridge, England: Cambridge University Press.
Smart, J. J. C. and Bernard Williams. (1973). *Utilitarianism: For and Against*. Cambridge, England: Cambridge University Press.
Sommerhoff, G. (1950) *Analytical Biology*. Oxford: Oxford University Press.
Spiegelberg, Herbert. (1982) *The Phenomenological Movement: A Historical Introduction*, 3rd revised edition. The Hague: Nijhoff.

Spillane, Robert. (1987) "Rhetoric as Remedy: Some Philosophical Antecedents of Psychotherapeutic Ethics," *British Journal of Medical Psychology*, 60, pp. 217–224.

Sterling, Peter and Joseph Eyer. (1981a) "Biological Basis of Stress-related Mortality," *Social Science and Medicine*, 15, pp. 3–42.

———. and Joseph Eyer. (1981b) "Allostasis: A New Paradigm to Explain Arousal Pathology," In *Handbook of Life Stress, Cognition and Health*, eds. S. Fischer and H. S. Reason. New York: John Wiley and Sons, pp. 629–649.

Stern, Donnel B. (1996) "The Social Construction of Therapeutic Action," *Psychoanalytic Inquiry*, 16, pp. 265–293.

Svenaeus, Fredrik. (1999) *The Hermeneutics of Medicine and the Phenomenology of Health*. Linköping, Sweden: Linköping Studies in Arts and Science, No 187.

———. (2001) *The Hermeneutics of Medicine and the Phenomenology of Health: Steps Towards a Philosophy of Medical Practice*, 2nd revised edition. Dordrecht: Kluwer Academic Publishers.

Tate, P. (1995) *The Doctor's Communication Handbook*. Oxford: Radcliffe Medical Press.

Taylor, A. E. (1960) *Plato: The Man and His Work*. London: Methuen.

Taylor, Charles. (1989) *Sources of the Self*. Cambridge, Mass.: Harvard University Press.

Tengland, Per-Anders. (1998) *Mental Health: A Philosophical Analysis*. Linköping, Sweden: Linköping Studies in Arts and Science, No 177.

———. (2001) *Mental Health: A Philosophical Analysis*, 2nd revised edition, Dordrecht: Kluwer Academic Publishers.

Tillich, Paul. (1952) *The Courage to Be*. New Haven: Yale University Press.

———. (1967) "The Meaning of Health in Religion and Medicine: Essays on Meaning, Value and Health." In *Religion and Medicine, Essays on Meaning, Values and Health*, ed. D. Belgum. Ames, Ia.: Iowa State University Press.

———. (1984) "The Relation of Religion and Health: Historical Considerations and Theoretical Questions." In *The Meaning of Health*, ed. P. LeFevre. Chicago: Exploration Press.

Tjeltveit, Alan C. (1999) *Ethics and Values in Psychotherapy*. London: Routledge & Kegan Paul.

Tomkins, Silvan S. (1982) "Affect theory." In *Emotion in the Human Face* (2nd edit.), ed. P. Ekman. New York: Cambridge University Press, pp. 353–95.

Toombs, S. Kay. (1992) *The Meaning of Illness. A Phenomenological Account of the Different Perspectives of Physician and Patient*. Dordrecht: Kluwer Academic Publishers.

Tornstam, Lars. (1997) "Gerotranscendence: The Contemplative Dimension of Aging," *Journal of Aging Studies*,11, 143–154.

Toulmin, Stephen. (1978) "Psychic Health, Mental Clarity, Self Knowledge and Other Virtues." In *Mental Health: Philosophical Perspectives*, eds. H. Tristram Engelhardt Jr. and Stuart F. Spicker. Dordrecht: D Reidel Publishing Company.

Wakefield, Jerome, C. (1992a) "The Concept of Mental Disorder: On the Boundary Between Biological Facts and Social Values," *American Psychologist*, 47, 373–388.

———. (1992b) "Disorder as Harmful Dysfunction: A Conceptual Critique of DSM-III-R's Definition of Mental Disorder," *Psychological Review*, 99, 232–247.

———. (1999a) "Evolutionary Versus Prototype Analyses of the Concept of Disorder," *Journal of Abnormal Psychology*, 108, 374–399.

———. (1999b) "Mental Disorder as a Black Box Essentialist Concept," *Journal of Abnormal Psychology*, 108, 465–472.

Whitbeck, Caroline. (1981) "A Theory of Health." In *Concepts of Health and Disease: Interdisciplinary Perspectives,* eds. A. L. Caplan, H. T. Engelhardt Jr., and J. J. McCartney. Reading, Massachusetts: Addison-Wesley, pp. 611-626.

WHO. (1948) *Official Records of the World Health Organization*, 2, p. 100.

Wiggins, D. (1987) *Needs, Values, Truth*. Oxford: Basil Blackwell.

Williams Alan. (1996) "Tydligare definitioner och bättre fakta krävs" [Clearer Definitions and Better Facts are Required], *Läkartidningen*, 93, pp. 3376–3379.

Winnicott, Donald W. (1984) *The Maturational Process and the Facilitating Environment*. New York: International Universities Press Inc.

Wright, Larry. (1976) *Teleological Explanations*. Berkeley: University of California Press.

Yalom, Irvin D. (1980) *Existential Psychotherapy*. New York: Basic Books, Inc.

Yamamoto, Keith R. (1985) "Steroid Receptor Regulated Transcription of Specific Genes and Gene Networks," *Annual Review of Genetics,* 19, pp. 209–252.

ABOUT THE CONTRIBUTORS

Jennifer Bullington, Ph.D., is a researcher at The National Institute for Psychosocial Factors and Health, Stockholm. Bullington has an academic background in philosophy (B.A.) and psychology (M.A.) and a clinical background as a physiotherapist and body-oriented psychotherapist. She defended her dissertation, *The Mysterious Life of the Body: A New Look at Psychosomatics,* in 1999 at the Department of Health and Society at the University of Linköping, Sweden. Her main areas of research interest are phenomenology and psychosomatic studies. She has participated in a number of empirical studies that employed a phenomenological method. Her current research concerns the care and treatment of psychosomatic patients with psychiatrics and general practice.

Olle Hellström, M.D., Ph.D., is a general practitioner at Vansbro Health Care Unit in the county of Dalarna, Sweden, where he participates in the education of doctors training to become family physicians. He is also a Research Fellow in the Center for Clinical Research in the city of Falun, Sweden. He has contributed to several anthologies about the interface between medicine and philosophy, published the book, *What Illness Will Tell*, and contributed richly to the *Journal of the Swedish Doctors' Association*. His main practical and research interest are doctor-patient relationships, dialogue-based medicine, and phenomenology.

Einar Jakobsson, Ph.D., is a psychologist, specialist in clinical psychology, and Senior Lecturer at Mälardalen University in Västerås and Eskilstuna, Sweden. In addition to psychotherapy research, he is at present carrying out research in children's health, professional intervention, and family cooperation in health care.

Per-Erik Liss, Ph.D., is Associate Professor at the Department of Health and Society, Linköping University. He has a background in philosophy and received his doctoral degree at the Department of Health and Society in 1990. He has a particular interest in the field of setting priorities in health care; his main publication in this field is *Health Care Need* (1993).

Nina Nikku, Ph.D., is Assistant Professor at the Tema Institute, Linköping University, and Lecturer at Jönköping University. She received her doctoral degree in 1997 from the Department of Health and Society, Linköping University, where she defended the thesis *Informative Paternalism: Studies in the Ethics of Promoting and Predicting Health*. Her recent research concerns ethics in the health care sector as well as the relation between ethics and politics.

Lennart Nordenfelt has been Professor of Philosophy of Medicine and Health Care at the Department of Health and Society, Linköping University, since 1987. He received his doctoral degree in theoretical philosophy at the University of Uppsala in 1974 and was subsequently appointed a lecturer in theoretical philosophy at the University of Stockholm. Nordenfelt has made contributions mainly to action theory, the theory of medicine, and the theory of health and welfare. His monographs include *On the Nature of Health* (1987 and 1995) and *Action, Ability, and Health* (2000).

Ingemar Nordin, Ph.D., is Professor of Philosophy at Linköping University's Department of Health and Society. His fields of research are the philosophy of science and technology, political philosophy, and the philosophy of medicine. He is especially interested in the problems of knowledge and rationality within medical research and practice. His main publications in the field are "The Role of Science in Medicine" in *Theoretical Medicine and Bioethics*, No. 3, Vol. 20, 1999, and "Expert and Non-Expert Knowledge in Medical Practise," in *Medicine, Health Care, and Philosophy*, No. 3, Vol. 3, 2000.

Bo Petersson is Professor of Practical Philosophy at Linköping University. He received his doctoral degree in practical philosophy at Uppsala University in 1973. His main research interests are within research ethics, moral philosophy, and the history of philosophy.

Bengt Richt, Ph.D., is Associate Professor at the Department of Health and Society, Linköping University, where he obtained his doctoral degree through a study of the living situation of juvenile diabetics and their families. His research interest is the sociology of health and illness, with special emphasis on the phenomenology and social psychology of chronic illness and on lay conceptions of health, illness, and health care, including complementary medicine. He has carried out research on elderly care and has recently published a book in Swedish about contact persons in elderly care. He is also involved in a study of the local implementation of a governmental program for health-promotion.

Per Sundström, born in 1955 in Stockholm, studied philology, philosophy, and theology at Uppsala University. He received his M.D. degree from the Karolinska Institute, Stockholm, and earned a Ph.D. in the philosophy of medicine at Linköping University. He is currently Professor of Medical Ethics on the Medical Faculty of the University of Oslo, Norway.

Fredrik Svenaeus, Ph.D., is Assistant Professor in Philosophy at Södertörn University College, Huddinge, Sweden. His main research interests are the phenomenology of medicine, philosophy of psychiatry, and bioethics. He defended his doctoral dissertation at the Department of Health and Society, Linköping University, in 1999. That thesis was recently published in a second, revised edition: *The Hermeneutics of Medicine and the Phenomenology of Health: Steps Towards a Philosophy of Medical Practice* (Kluwer, 2001).

Per-Anders Tengland, Ph.D., is a Senior Lecturer at the Department of Public Health at Malmö University, Sweden. He received his Ph.D. from the Department of Health and Society at Linköping University. Kluwer Academic Publishers has published a revised version of his doctoral thesis under the title *Mental Health: A Philosophical Analysis* (2001). Lately Tengland has taken an interest in the theoretical foundations of psychotherapy, writing about the concepts of "congruence" and "empathy."

Kristin Zeiler, M.A. (in medicine and religious sciences), is a Ph.D. candidate at the Department of Health and Society, Linköping University. Her study *Lifeviews, Human Nature, and Genetics* aims at exploring moral questions as discussed by gynecologists and geneticists in Sweden, Italy, and the United Kingdom in the context of preimplantation genetic diagnosis and germ-line gene therapy. Her main areas of interest are bioethics, feminist ethics, global/ecumenical ethics, and research on life views. The research reported in her essay was financially supported by the Swedish Foundation for International Cooperation in Research and Higher Education (STINT).

INDEX

VIBS

The **Value Inquiry Book Series** is co-sponsored by:

Adler School of Professional Psychology
American Indian Philosophy Association
American Maritain Association
American Society for Value Inquiry
Association for Process Philosophy of Education
Canadian Society for Philosophical Practice
Center for Bioethics, University of Turku
Center for Professional and Applied Ethics, University of North Carolina at Charlotte
Center for Research in Cognitive Science, Autonomous University of Barcelona
Centre for Applied Ethics, Hong Kong Baptist University
Centre for Cultural Research, Aarhus University
Centre for Professional Ethics, University of Central Lancashire
Centre for the Study of Philosophy and Religion, University College of Cape Breton
College of Education and Allied Professions, Bowling Green State University
College of Liberal Arts, Rochester Institute of Technology
Concerned Philosophers for Peace
Conference of Philosophical Societies
Department of Moral and Social Philosophy, University of Helsinki
Gannon University
Gilson Society
Ikeda University
Institute of Philosophy of the High Council of Scientific Research, Spain
International Academy of Philosophy of the Principality of Liechtenstein
International Association of Bioethics
International Center for the Arts, Humanities, and Value Inquiry
International Society for Universal Dialogue
Natural Law Society
Personalist Discussion Group
Philosophical Society of Finland
Philosophy Born of Struggle Association
Philosophy Seminar, University of Mainz
Pragmatism Archive
R.S. Hartman Institute for Formal and Applied Axiology
Research Institute, Lakeridge Health Corporation
Russian Philosophical Society
Society for Iberian and Latin-American Thought
Society for the Philosophic Study of Genocide and the Holocaust
Society for the Philosophy of Sex and Love
Yves R. Simon Institute

Titles Published

1. Noel Balzer, *The Human Being as a Logical Thinker*

2. Archie J. Bahm, *Axiology: The Science of Values*

3. H. P. P. (Hennie) Lötter, *Justice for an Unjust Society*

4. H. G. Callaway, *Context for Meaning and Analysis: A Critical Study in the Philosophy of Language*

5. Benjamin S. Llamzon, *A Humane Case for Moral Intuition*

6. James R. Watson, *Between Auschwitz and Tradition: Postmodern Reflections on the Task of Thinking*. A volume in **Holocaust and Genocide Studies**

7. Robert S. Hartman, *Freedom to Live: The Robert Hartman Story, edited by Arthur R. Ellis*. A volume in **Hartman Institute Axiology Studies**

8. Archie J. Bahm, *Ethics: The Science of Oughtness*

9. George David Miller, *An Idiosyncratic Ethics; Or, the Lauramachean Ethics*

10. Joseph P. DeMarco, *A Coherence Theory in Ethics*

11. Frank G. Forrest, *Valuemetrics*℠: *The Science of Personal and Professional Ethics*. A volume in **Hartman Institute Axiology Studies**

12. William Gerber, *The Meaning of Life: Insights of the World's Great Thinkers*

13. Richard T. Hull, Editor, *A Quarter Century of Value Inquiry: Presidential Addresses of the American Society for Value Inquiry*. A volume in **Histories and Addresses of Philosophical Societies**

14. William Gerber, *Nuggets of Wisdom from Great Jewish Thinkers: From Biblical Times to the Present*

15. Sidney Axinn, *The Logic of Hope: Extensions of Kant's View of Religion*

16. Messay Kebede, *Meaning and Development*

17. Amihud Gilead, *The Platonic Odyssey: A Philosophical-Literary Inquiry into the Phaedo*

18. Necip Fikri Alican, *Mill's Principle of Utility: A Defense of John Stuart Mill's Notorious Proof.* A volume in **Universal Justice**

19. Michael H. Mitias, Editor, *Philosophy and Architecture.*

20. Roger T. Simonds, *Rational Individualism: The Perennial Philosophy of Legal Interpretation.* A volume in **Natural Law Studies**

21. William Pencak, *The Conflict of Law and Justice in the Icelandic Sagas*

22. Samuel M. Natale and Brian M. Rothschild, Editors, *Values, Work, Education: The Meanings of Work*

23. N. Georgopoulos and Michael Heim, Editors, *Being Human in the Ultimate: Studies in the Thought of John M. Anderson*

24. Robert Wesson and Patricia A. Williams, Editors, *Evolution and Human Values*

25. Wim J. van der Steen, *Facts, Values, and Methodology: A New Approach to Ethics*

26. Avi Sagi and Daniel Statman, *Religion and Morality*

27. Albert William Levi, *The High Road of Humanity: The Seven Ethical Ages of Western Man*, Edited by Donald Phillip Verene and Molly Black Verene

28. Samuel M. Natale and Brian M. Rothschild, Editors, *Work Values: Education, Organization, and Religious Concerns*

29. Laurence F. Bove and Laura Duhan Kaplan, Editors, From the Eye of the Storm: Regional Conflicts and the Philosophy of Peace. A volume in **Philosophy of Peace**

30. Robin Attfield, *Value, Obligation, and Meta-Ethics*

31. William Gerber, *The Deepest Questions You Can Ask About God: As Answered by the World's Great Thinkers*

32. Daniel Statman, *Moral Dilemmas*

33. Rem B. Edwards, Editor, *Formal Axiology and Its Critics.* A volume in **Hartman Institute Axiology Studies**

34. George David Miller and Conrad P. Pritscher, *On Education and Values: In Praise of Pariahs and Nomads.* A volume in **Philosophy of Education**

35. Paul S. Penner, *Altruistic Behavior: An Inquiry into Motivation*

36. Corbin Fowler, *Morality for Moderns*

37. Giambattista Vico, *The Art of Rhetoric* (*Institutiones Oratoriae*, 1711–1741), from the definitive Latin text and notes, Italian commentary and introduction by Giuliano Crifò. Translated and Edited by Giorgio A. Pinton and Arthur W. Shippee. A volume in **Values in Italian Philosophy**

38. W. H. Werkmeister, *Martin Heidegger on the Way*. Edited by Richard T. Hull A volume in **Werkmeister Studies**

39. Phillip Stambovsky, *Myth and the Limits of Reason*

40. Samantha Brennan, Tracy Isaacs, and Michael Milde, Editors, *A Question of Values: New Canadian Perspectives in Ethics and Political Philosophy*

41. Peter A. Redpath, *Cartesian Nightmare: An Introduction to Transcendental Sophistry*. A volume in **Studies in the History of Western Philosophy**

42. Clark Butler, *History as the Story of Freedom: Philosophy in Intercultural Context*, with responses by sixteen scholars

43. Dennis Rohatyn, *Philosophy History Sophistry*

44. Leon Shaskolsky Sheleff, *Social Cohesion and Legal Coercion: A Critique of Weber, Durkheim, and Marx*. Afterword by Virginia Black

45. Alan Soble, Editor, *Sex, Love, and Friendship: Studies of the Society for the Philosophy of Sex and Love, 1977–1992*. A volume **in Histories and Addresses of Philosophical Societies**

46. Peter A. Redpath, *Wisdom's Odyssey: From Philosophy to Transcendental Sophistry*. A volume in **Studies in the History of Western Philosophy**

47. Albert A. Anderson, *Universal Justice: A Dialectical Approach*. A volume in **Universal Justice**

48. Pio Colonnello, *The Philosophy of José Gaos*. Translated from Italian by Peter Cocozzella. Edited by Myra Moss. Introduction by Giovanni Gullace. A volume in **Values in Italian Philosophy**

49. Laura Duhan Kaplan and Laurence F. Bove, Editors, *Philosophical Perspectives on Power and Domination: Theories and Practices*. A volume in **Philosophy of Peace**

50. Gregory F. Mellema, *Collective Responsibility*

51. Josef Seifert, *What Is Life? The Originality, Irreducibility, and Value of Life.* A volume in **Central-European Value Studies**

52. William Gerber, *Anatomy of What We Value Most*

53. Armando Molina, *Our Ways: Values and Character*, Edited by Rem B. Edwards. A volume in **Hartman Institute Axiology Studies**

54. Kathleen J. Wininger, *Nietzsche's Reclamation of Philosophy.* A volume in **Central-European Value Studies**

55. Thomas Magnell, Editor, *Explorations of Value*

56. HPP (Hennie) Lötter, *Injustice, Violence, and Peace: The Case of South Africa.* A volume in **Philosophy of Peace**

57. Lennart Nordenfelt, *Talking About Health: A Philosophical Dialogue.* A volume in **Nordic Value Studies**

58. Jon Mills and Janusz A. Polanowski, *The Ontology of Prejudice.* A volume in **Philosophy and Psychology**

59. Leena Vilkka, *The Intrinsic Value of Nature*

60. Palmer Talbutt, Jr., Rough Dialectics: *Sorokin's Philosophy of Value*, with contributions by Lawrence T. Nichols and Pitirim A. Sorokin

61. C. L. Sheng, *A Utilitarian General Theory of Value*

62. George David Miller, *Negotiating Toward Truth: The Extinction of Teachers and Students.* Epilogue by Mark Roelof Eleveld. A volume in **Philosophy of Education**

63. William Gerber, *Love, Poetry, and Immortality: Luminous Insights of the World's Great Thinkers*

64. Dane R. Gordon, Editor, *Philosophy in Post-Communist Europe.* A volume in **Post-Communist European Thought**

65. Dane R. Gordon and Józef Niznik, Editors, *Criticism and Defense of Rationality in Contemporary Philosophy.* A volume in **Post-Communist European Thought**

66. John R. Shook, *Pragmatism: An Annotated Bibliography, 1898-1940.* With contributions by E. Paul Colella, Lesley Friedman, Frank X. Ryan, and Ignas K. Skrupskelis

67. Lansana Keita, *The Human Project and the Temptations of Science*

68. Michael M. Kazanjian, *Phenomenology and Education: Cosmology, Co-Being, and Core Curriculum.* A volume in **Philosophy of Education**

69. James W. Vice, *The Reopening of the American Mind: On Skepticism and Constitutionalism*

70. Sarah Bishop Merrill, *Defining Personhood: Toward the Ethics of Quality in Clinical Care*

71. Dane R. Gordon, *Philosophy and Vision*

72. Alan Milchman and Alan Rosenberg, Editors, *Postmodernism and the Holocaust.* A volume in **Holocaust and Genocide Studies**

73. Peter A. Redpath, *Masquerade of the Dream Walkers: Prophetic Theology from the Cartesians to Hegel.* A volume in **Studies in the History of Western Philosophy**

74. Malcolm D. Evans, *Whitehead and Philosophy of Education: The Seamless Coat of Learning.* A volume in **Philosophy of Education**

75. Warren E. Steinkraus, *Taking Religious Claims Seriously: A Philosophy of Religion*, edited by Michael H. Mitias. A volume in **Universal Justice**

76. Thomas Magnell, Editor, *Values and Education*

77. Kenneth A. Bryson, *Persons and Immortality.* A volume in **Natural Law Studies**

78. Steven V. Hicks, *International Law and the Possibility of a Just World Order: An Essay on Hegel's Universalism.* A volume in **Universal Justice**

79. E. F. Kaelin, *Texts on Texts and Textuality: A Phenomenology of Literary Art*, Edited by Ellen J. Burns

80. Amihud Gilead, Saving Possibilities: A Study in Philosophical Psychology. *A volume in Philosophy and Psychology*

81. André Mineau, *The Making of the Holocaust: Ideology and Ethics in the Systems Perspective.* A volume in **Holocaust and Genocide Studies**

82. Howard P. Kainz, *Politically Incorrect Dialogues: Topics Not Discussed in Polite Circles*

83. Veikko Launis, Juhani Pietarinen, and Juha Räikkä, Editors, *Genes and Morality: New Essays*. A volume in **Nordic Value Studies**

84. Steven Schroeder, *The Metaphysics of Cooperation: A Study of F. D. Maurice*

85. Caroline Joan ("Kay") S. Picart, *Thomas Mann and Friedrich Nietzsche: Eroticism, Death, Music, and Laughter*. A volume in **Central-European Value Studies**

86. G. John M. Abbarno, Editor, *The Ethics of Homelessness: Philosophical Perspectives*

87. James Giles, Editor, *French Existentialism: Consciousness, Ethics, and Relations with Others*. A volume in **Nordic Value Studies**

88. Deane Curtin and Robert Litke, Editors, *Institutional Violence*. A volume in **Philosophy of Peace**

89. Yuval Lurie, *Cultural Beings: Reading the Philosophers of Genesis*

90. Sandra A. Wawrytko, Editor. *The Problem of Evil: An Intercultural Exploration*. A volume in **Philosophy and Psychology**

91. Gary J. Acquaviva, *Values, Violence, and Our Future*. A volume in **Hartman Institute Axiology Studies**

92. Michael R. Rhodes, *Coercion: A Nonevaluative Approach*

93. Jacques Kriel, *Matter, Mind, and Medicine: Transforming the Clinical Method*

94. Haim Gordon, *Dwelling Poetically: Educational Challenges in Heidegger's Thinking on Poetry*. A volume in **Philosophy of Education**

95. Ludwig Grünberg, *The Mystery of Values: Studies in Axiology*, edited by Cornelia Grünberg and Laura Grünberg

96. Gerhold K. Becker, Editor, *The Moral Status of Persons: Perspectives on Bioethics*. A volume in **Studies in Applied Ethics**

97. Roxanne Claire Farrar, *Sartrean Dialectics: A Method for Critical Discourse on Aesthetic Experience*

98. Ugo Spirito, *Memoirs of the Twentieth Century*. Translated from Italian and edited by Anthony G. Costantini. A volume in **Values in Italian Philosophy**

99. Steven Schroeder, *Between Freedom and Necessity: An Essay on the Place of Value*

100. Foster N. Walker, *Enjoyment and the Activity of Mind: Dialogues on Whitehead and Education*. A volume in **Philosophy of Education**

101. Avi Sagi, Kierkegaard, *Religion, and Existence: The Voyage of the Self*. Translated from Hebrew by Batya Stein

102. Bennie R. Crockett, Jr., Editor, *Addresses of the Mississippi Philosophical Association*. A volume in **Histories and Addresses of Philosophical Societies**

103. Paul van Dijk, *Anthropology in the Age of Technology: The Philosophical Contribution of Günther Anders*

104. Giambattista Vico, *Universal Right*. Translated from Latin and Edited by Giorgio Pinton and Margaret Diehl. A volume in **Values in Italian Philosophy**

105. Judith Presler and Sally J. Scholz, Editors, *Peacemaking: Lessons from the Past, Visions for the Future*. A volume in **Philosophy of Peace**

106. Dennis Bonnette, *Origin of the Human Species*. A volume in **Studies in the History of Western Philosophy**

107. Phyllis Chiasson, *Peirce's Pragmatism: The Design for Thinking*. A volume in **Studies in Pragmatism and Values**

108. Dan Stone, Editor, *Theoretical Interpretations of the Ho*locaust. A volume in **Holocaust and Genocide Studies**

109. Raymond Angelo Belliotti, *What Is the Meaning of Human Life?*

110. Lennart Nordenfelt, *Health, Science, and Ordinary Language*, with Contributions by George Khushf and K. W. M. Fulford

111. Daryl Koehn, *Local Insights, Global Ethics for Business*. A volume in **Studies in Applied Ethics**

112. Matti Häyry and Tuija Takala, Editors, *The Future of Value Inquiry*. A volume in **Nordic Value Studies**

113. Conrad P. Pritscher, *Quantum Learning: Beyond Duality*

114. Thomas M. Dicken and Rem B. Edwards. *Dialogues on Values and Centers of Value: Old Friends, New Thoughts*. A volume in **Hartman Institute Axiology Studies**

115. Rem B. Edwards, *What Caused the Big Bang?* A volume in **Philosophy and Religion**

116. Jon Mills, Editor, *A Pedagogy of Becoming*. A volume in **Philosophy of Education**

117. Robert T. Radford, *Cicero: A Study in the Origins of Republican Philosophy*. A volume in **Studies in the History of Western Philosophy**

118. Arleen L. F. Salles and María Julia Bertomeu, Editors, *Bioethics: Latin American Perspectives*. A volume in **Philosophy in Latin America**

119. Nicola Abbagnano, *The Human Project: The Year 2000*, with an Interview by Guiseppe Grieco. Translated from Italian by Bruno Martini and Nino Langiulli. Edited with an introduction by Nino Langiulli. A volume in **Studies in the History of Western Philosophy**

120. Daniel M. Haybron, Editor, *Earth's Abominations: Philosophical Studies of Evil*. A volume in **Personalist Studies**

121. Anna T. Challenger, *Philosophy and Art in Gurdjieff's Beelzebub: A Modern Sufi Odyssey*

122. George David Miller, *Peace, Value, and Wisdom: The Educational Philosophy of Daisaku Ikeda*. A volume in **Daisaku Ikeda Studies**

123. Haim Gordon and Rivca Gordon, *Sophistry and Twentieth-Century Art*

124. Thomas O Buford and Harold H. Oliver, Editors *Personalism Revisited: Its Proponents and Critics*. A volume in **Histories and Addresses of Philosophical Societies**

125. Avi Sagi, *Albert Camus and the Philosophy of the Absurd*. Translated from Hebrew by Batya Stein

126. Robert S. Hartman, *The Knowledge of Good: Critique of Axiologic l Reason*. Expanded translation from the Spanish by Robert S. Hartman. Edited by Arthur R. Ellis and Rem B. Edwards. A volume in **Hartman Institute Axiology Studies**

127. Alison Bailey and Paula J. Smithka, Editors. *Community, Diversity, and Difference: Implications for Peace*. A volume in **Philosophy of Peace**

128. Oscar Vilarroya, *The Dissolution of Mind: A Fable of How Experience Gives Rise to Cognition*. A volume in **Cognitive Science**

129. Paul Bube and Jefferey Geller, Editors, *Conversations with Pragmatism*. A volume in **Studies in Pragmatism and Values**

130. Richard Rumana, *Richard Rorty: An Annotated Bibliography of Secondary Literature*. A volume in **Studies in Pragmatism and Values**

131. Stephen Schneck, *Max Scheler's Acting Persons: New Perspectives*. A volume in **Personalist Studies**

132. Michael Kazanjian, *Learning Values Lifelong: From Inert Ideas to Wholes*. A volume in **Philosophy of Education**

133. Rudolph Alexander Kofi Cain: *Alain Leroy Locke: Race, Culture, and the Education of African American Adults*. A Volume in **African American Philosophy.**

134. Werner J. Krieglstein, *Compassion: A New Philosophy of the Other*.

135. Robert N. Fisher, Daniel T. Primozic, Peter A. Day, and Joel A. Thompson, Editors, *Suffering, Death, and Identity*. A volume in **Personalist Studies**

136. Steven Schroeder, *Touching Philosophy, Sounding Religion, Placing Education*. A volume in **Philosophy of Education.**

137. Guy DeBrock, *Process Pragmatism: Essays on a Quiet Philosophical Revolution*. **A volume in Studies in Pragmatism and Values**

138. Lennart Nordenfelt and Per-Erik Liss, Editors, *Dimensions of Health and Health Promotion*